High Entropy Alloys - Composition and Microstructure Design

Edited by Yu Yin, Han Huang, Mingxing Zhang and Libo Zhou

Published in London, United Kingdom

High Entropy Alloys - Composition and Microstructure Design
http://dx.doi.org/10.5772/intechopen.1001618
Edited by Yu Yin, Han Huang, Mingxing Zhang and Libo Zhou

Contributors
Alexandre Melhorance Barboza, Amir H. Naghdi, Dario Massa, Dekui Mu, Han Huang, Huapan Xiao, Ivan Napoleão Bastos, Jens Hänisch, Jike Wang, Junyu Zhang, Kamran Karimi, Libo Zhou, Loena Marins de Couto, Luis César R. Aliaga, Mingxing Zhang, Min Wei, Pablo Cayado, Stefanos Papanikolaou, Yu Yin

First published in London, United Kingdom, 2025 by IntechOpen
IntechOpen is the global imprint of INTECHOPEN LIMITED, registered in England and Wales, registration number: 11086078, 167-169 Great Portland Street, London, W1W 5PF, United Kingdom

For EU product safety concerns: IN TECH d.o.o., Prolaz Marije Krucifikse Kozulić 3, 51000 Rijeka, Croatia, info@intechopen.com or visit our website at intechopen.com.

British Library Cataloguing-in-Publication Data
A catalogue record for this book is available from the British Library

High Entropy Alloys - Composition and Microstructure Design
Edited by Yu Yin, Han Huang, Mingxing Zhang and Libo Zhou
p. cm.

This title is part of the Materials Science Book Series, Volume 3
Topic: Composite Materials
Series Editor: Chonghe Li
Topic Editor: Anna Boczkowska
Assocaite Topic Editors: Kamil Dydek and Bartlomiej Przybyszewski

Print ISBN 978-0-85466-299-9
Online ISBN 978-0-85466-298-2
eBook (PDF) ISBN 978-0-85466-300-2
ISSN 3049-8856

If disposing of this product, please recycle the paper responsibly.

IntechOpen Book Series

Materials Science

Volume 3

Aims and Scope of the Series

Materials science has always occupied an extremely high position in the human development process. As we explore the oceans of stars, various industries have put forward more stringent requirements for the performance of materials, forcing us to pay more and more attention to the development of new materials. At the same time, the formation of a data-driven scientific paradigm is dramatically shortening the development cycle of new materials. The huge data generated by synergistically combining theories, high-throughput experiments, high-throughput computation, and artificial intelligence is greatly contributing to our ability to utilize materials science to solve real-world problems. The three topics of this book series - Metals and Nonmetals; Composite Materials; and Surface Science - will address important areas of advancement in materials science. There will be a range of interesting works published under these topics.

Meet the Series Editor

Prof. Chonghe Li received his Ph.D. from the Chinese Academy of Sciences in 1995. From 1995 to 2000, he worked as a researcher at the Shanghai Institute of Metallurgy, Chinese Academy of Sciences, where he also served as director of the research laboratory. In 2000, he was appointed Professor at the Institute of High-Performance Computing in Singapore, where he worked on computation and simulation in materials science until 2004. Since then, he has been a professor at the School of Materials Science and Engineering, Shanghai University, China, as well as the director of the Shanghai Specialty Casting Engineering and Technology Research Center. Prof. Li's research focuses on titanium alloys, titanium-aluminum single crystals, intermetallic compounds, theoretical calculations, alloy design, and special refractory materials. His broad scientific expertise is well recognized by the scientific community around the world. He is a member of the editorial board of the journal Metals. As an author, he has published more than 200 peer-reviewed papers, 2 books, and over 40 patents.

Meet the Volume Editors

Dr. Yu Yin is an Associate Professor at Sun Yat-sen University in China, specializing in advanced manufacturing technologies for high-performance alloys. With over 50 publications in international journals such as materials science and additive manufacturing, his work has garnered over 2,300 citations and an H-index of 26. In 2023, he was nominated to the Sigma Xi Scientific Research Honor Society. He serves as a Youth Editorial Board Member for the *Journal of Materials Science & Technology* and *Tungsten* and as a reviewer for top journals like Acta Materialia and Scripta Materialia. Dr. Yin is recognized for his contributions to alloy design, microstructure control, and advanced manufacturing techniques.

Professor Zhang obtained his Bachelor of Engineering from the Inner Mongolian University of Science and Technology and Master of Engineering from Northwestern Polytechnical University, China. In 1997 he was awarded his Ph.D. degree by The University of Queensland. Professor Mingxing Zhang's research interests are in the additive manufacturing of metals and MAX phase materials, high entropy alloys, new alloy design through machine learning and application of crystallography to engineering materials, surface engineering of metals, and grain refinement for cast metals. His major contributions include the invention of the edge-to-edge matching crystallographic model, and successful application of this model to grain refinement to cast metals for seeking new and more effective grain refiners, and to metal additive manufacturing for increasing the 3D printability.

Han Huang is the Chair Professor at Sun Yat-sen University's School of Advanced Manufacturing, China, and Emeritus Professor at The University of Queensland, Australia. He holds a Ph.D. from The University of Western Australia and ME/BE degrees from Huazhong University of Science and Technology. Prof. Huang has published over 300 SCI journal articles, with 14,450+ citations and an h-index 66. He received accolades such as Australia Research Council Future Fellow, Queensland International Fellow, and Singapore National Technology Award. He is a Fellow of the International Society of Nanomanufacturing and serves as Associate Editor for journals like the International Journal of Mechanical Sciences and the International Journal of Extreme Manufacturing. His research focuses on ultraprecision machining, additive manufacturing, and intelligent manufacturing.

Prof. Libo Zhou earned his MS and Ph.D. degrees from Tohoku University, Japan, in 1988 and 1991, respectively. With over 30 years of experience in R&D of ultra-precision machining technologies, he specializes in the machining and testing semiconductor materials. His work is recognized for its originality, foresight, and profound academic contributions. His research interests cover traditional and non-traditional manufacturing processes, which have recently been extended to applications of AI in smart manufacturing. His major contributions include material removal mechanisms beyond plastic wave propagation rate, developing an integrated grinding system for large-scale Si wafers, developing a CMG wheel and defect-free process, and developing AI applications for intelligent manufacturing.

Contents

Preface

High-entropy alloys (HEAs) have gained significant attention for their remarkable properties, such as enhanced strength, corrosion resistance, and fracture toughness. This book, *High Entropy Alloys – Composition and Microstructure Design*, brings together key insights and cutting-edge developments in the field, offering a comprehensive exploration of the design and application of HEAs.

Chapter 1 briefly reviews the composition design of HEAs, summarizing the various strategies employed over the years, including physical modeling, computational approaches, and self-optimizing strategies. In Chapter 2, the focus shifts to the mechanical properties of HEAs, with an emphasis on the role of machine learning tools in bridging the gap between composition, microstructure, and mechanical response. Chapter 3 delves into the exciting field of HEA catalysts, examining how machine learning aids in discovering high-performance catalysts and unveiling catalytic reaction mechanisms. Chapter 4 explores the use of molecular dynamics simulations to investigate the mechanical properties and phase transitions in the Hf-Nb-Ta-Ti-Zr HEA, offering a deeper understanding of its microscopic behavior. Finally, Chapter 5 reviews the emerging field of high-entropy superconductors, highlighting their potential and the latest advancements in synthesis and characterization.

This volume is a valuable resource for materials science researchers and practitioners, offering theoretical and practical insights into the design and application of HEAs. We want to extend our gratitude to those who contributed to the editorial process and helped bring this work to fruition.

Yu Yin, Han Huang and Libo Zhou
School of Advanced Manufacturing,
Sun Yat-sen University,
Shenzhen, China

Mingxing Zhang
School of Mechanical and Mining Engineering,
University of Queensland,
Brisbane, Australia

Chapter 1

Composition Design of High-Entropy Alloys: A Brief Review

Yu Yin, Libo Zhou, Dekui Mu, Han Huang, Mingxing Zhang and Huapan Xiao

Abstract

High-entropy alloys (HEAs) have attracted significant interest since their conceptualization in 2004, owing to their exceptional properties such as high strength, low-temperature fracture toughness and corrosion resistance. HEAs typically comprise solid solution phases, intermetallics, and/or amorphous phases, with solid solutions being particularly desired for their superior mechanical properties. Accurately predicting phase constituents in HEAs remains a formidable challenge due to their infinite composition space. During the past decade, various strategies have been proposed to design the composition of HEAs. This review aims to provide a brief overview of these strategies, including physical modeling, computational approaches and self-optimizing strategies, thereby providing researchers with current insights and knowledge in this rapidly advancing field.

Keywords: high-entropy alloys, composition design, phase prediction, physical modeling, phase diagram calculation, first-principles calculations, machine learning, self-optimizing strategy

1. Introduction

High-entropy alloys (HEAs), also known as multicomponent alloys or multi-principal element alloys (MPEAs), were first proposed by Jien-Wei Yeh and Brian Cantor in the same year of 2004 [1–3]. HEAs are initially defined as those composed of over five principal elements in equiatomic ratios [1]. Due to the high mixing entropy, single-phase solid-solution (SS) structures tend to form in HEAs, including the FCC (face-centred cubic), BCC (body-centred cubic) and HCP (hexagonal close-packed) phases [4–9]. Recently, the definition of HEAs has been extended to those with a composition of each element between 5 and 35 at.% [6, 10]. Moreover, HEA is not strictly limited to single-phase SS alloys with over five principal elements. To avoid confusion, in this review, the term "HEAs" encompasses both single-phase and multi-phase alloys with equiatomic or non-equiatomic compositions involving multiple principal elements. At present, certain equiatomic quaternary alloys and non-equiatomic multicomponent alloys with multi-phase structures are also classified as HEAs. Examples include NbMoTaW,

$Fe_{50}Mn_{30}Co_{10}Cr_{10}$, $Ta_{0.6}HfZrTi$, TiZrHfNb and CoCrFeNi alloys [11–17]. These alloys with multi-phase structures are often referred to as complex concentrated alloys or compositionally complicated alloys (CCAs) [11, 17–19].

Up to now, more than 400 HEA systems with diverse compositions have been reported. Researchers have shown considerable interest in developing HEAs with either single-phase SS structures or SS matrix containing nano-sized intermetallic phases due to their exceptional properties. These SS structures commonly adopt FCC, BCC, or HCP crystal structures [4–9]. The mechanical properties of HEAs are primarily dictated by their phase constituents and microstructures. For example, BCC-structured HEAs tend to exhibit high strength but low ductility, whereas FCC-structured HEAs demonstrate high ductility with relatively lower strength [20, 21]. In addition, HEAs combining disordered FCC with coherent intermetallic phases display extraordinary mechanical properties. A notable example is the 6-component $Ni_{32.8}Fe_{21.9}Co_{21.9}Cr_{10.9}Al_{7.5}Ti_{5.0}$ HEA, which features a dual-phase nanoscale lamellar architecture structure comprising FCC and $L1_2$ phases. This alloy exhibits exceptional strength (yield strength >2 GPa) and ductility (uniform elongation >16%) [22]. Therefore, accurately predicting phase constituents in HEAs holds significant implications both theoretically and for practical applications [23]. However, designing HEAs with desired phase constituents remains a formidable challenge due to the vast compositional space of these alloys. To tackle this challenge, various strategies have emerged, including physical modeling, computational methods and self-optimizing strategies [23–26]. The following sections briefly summarize these methods, providing researchers with updated insights and knowledge in this field.

2. Physical model methods

Currently, at least eight physical models have been proposed for predicting the phase structure of high-entropy alloys. These models can be categorized into two groups: parameter-based and free energy-based models [24]. Here, we primarily introduce the most commonly used parameter rules due to their simplicity and relatively high accuracy. These include the ΔH_{mix} (mixing enthalpy) and δ (difference in atomic radius) rules [27–29], the Ω (the ratio of mixing entropy to mixing enthalpy) and δ rule [27], and the VEC (valence electron concentration) rule [6, 30, 31]. The calculation methods and criteria for phase prediction are discussed in the following sections.

At present, accurately determining ΔG for multi-element systems for specific compositions and temperatures is challenging. To address this, Takeuchi and Inoue [32] proposed a hypothesis: ΔG at a given composition is proportional to the ΔG_{mix} (mixing free energy) of the liquid phase, which can be expressed as the following equation:

$$\Delta G_{mix} = \Delta H_{mix} - T\Delta S_{mix} \tag{1}$$

where ΔH_{mix} is the mixing enthalpy, T is the absolute temperature, ΔS_{mix} is the mixing entropy. The mixing enthalpy of a multicomponent alloy with n-elements can be expressed as eq. (2) [32]:

$$\Delta H_{mix} = \sum_{i=1,\, i\neq j}^{n} \Omega_{ij} c_i c_j \tag{2}$$

where c_i or c_j is the atomic percentage of the ith and jth component, and $\Omega_{ij} = 4\Delta H_{AB}^{mix}$ is the regular solution interaction parameter between the ith and jth elements. The values of ΔH_{AB}^{mix} AB (mixing enthalpy of binary liquid alloys) can be obtained in Ref. [33].

Based on Boltzmann's hypothesis, the mixing entropy of solution n-elements with can be expressed as follows:

$$\Delta S_{mix} = -R\sum_{i=1}^{n} c_i \ln c_i \tag{3}$$

where c_i is mole percent of component, $\sum_{i=1}^{n} c_i = 1$, and $R(= 8.314JK^{-1}mol^{-1})$ is gas constant. For alloys with equiatomic ratios, the mixing entropy is the maximum. Thus, equiatomic or near-equiatomic HEAs generally show higher mixing entropy than their conventional counterparts.

According to Sheng et al. [31], the formation of solid-solution phases in HEAs can be assessed by comparing the values of ΔH_{mix} and ΔS_{mix}. But, the ΔH_{mix} and ΔS_{mix} values for HEAs with different phase constituents often significantly overlap. To improve phase prediction accuracy, the atomic size mismatch effect of constituent elements has been considered [34, 35]. This is justified by two main reasons. First, large differences in atomic size ratios cause substantial lattice distortion in HEAs, increasing strain energy. Such significant atomic radius differences elevate the free energy of alloys, destabilizing solid solutions. Second, large atomic size disparities can raise diffusion activation energy, leading to sluggish diffusion in alloys. Consequently, phase transformation rate decreases, potentially causing elemental segregation, nano-sized precipitates, or even amorphous phases. Therefore, to comprehensively describe the impact of atomic radius differences in HEAs, the parameter δ is expressed as follows:

$$\delta = \sqrt{\sum_{i=1}^{n} c_i\left(1 - \frac{r_i}{\bar{r}}\right)^2} \tag{4}$$

where c_i is the molar ratio of the ith component, $\bar{r} = \sum_{i=1}^{n} c_i r_i$ is the average atomic radius and r_i is the atomic radius. In the calculation, the values of $r_i$can be obtained in the Ref. [36]. According to the ΔH_{mix}-δ rule and **Figure 1** [31, 34, 35], solid solution

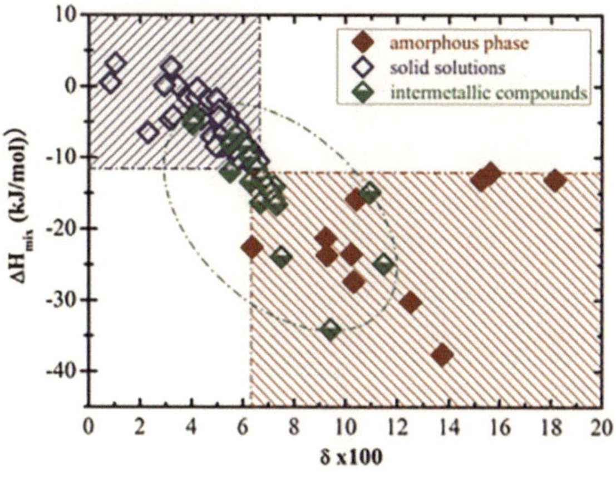

Figure 1.
A δ-ΔH$_{mix}$ plot illustrating the phase selection in HEAs [34].

phases usually form when ΔH_{mix} and δ (%) are in a range around: −11.6 KJ/mol $\leq \Delta H_{mix} \leq$ 3.2 KJ/mol and $\delta(\%) \leq 6.6$ [34].

Besides, the Ω rule developed by Zhang and co-workers [27] integrates another two parameters, including ΔS_{mix} (mixing entropy) and T_m (melting temperature of n-element alloy), which can be expressed as follows:

$$\Omega = \frac{T_m \Delta S_{mix}}{|\Delta H_{mix}|} \tag{5}$$

Disregarding the solid-state phase transitions, phase formation generally takes place near the alloy's melting temperature (T_m). Consequently, T_m is employed for the entropy term $T\Delta S_{mix}$. The melting temperature of n-elements alloys, T_m, is determined using the rule of mixtures:

$$T_m = \sum_{i=1}^{n} c_i (T_m)_i \tag{6}$$

where $(T_m)_i$ is the melting point of the ith component of alloy.

According to eqs. (4) and (5), Zhang et al. proposed a novel solid-solution formation rule for HEAs through the calculation of parameters Ω and δ for equiatomic or near-equiatomic HEAs. According to **Figure 2**, Ω≥1.1, $\delta \leq 6.6\%$ is the rough criteria for the formation of SS phase in HEAs [27].

Currently, two-parameter rules (ΔH_{mix} and δ, Ω and δ) are widely used for predicting phase types such as solid solution phases, intermetallics, and amorphous phases in high-entropy alloys (HEAs), due to their simplicity and relatively high accuracy. However, another parameter is necessary for predicting the specific phase structure, such as FCC, BCC, or hexagonal close-packed (HCP). According to Guo et al. [31] and Battezzati et al. [37], the valence electron concentration (VEC) of HEAs is another critical parameter influencing phase formation. The average VEC of an alloy can be calculated using Eq. 7, where $(VEC)_i$ is the VEC for the ith element [31].

$$VEC = \sum\nolimits_{i=1}^{n} c_i (VEC)_i \tag{7}$$

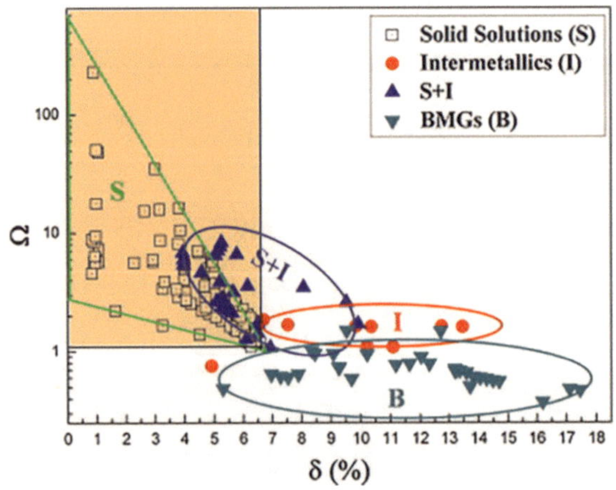

Figure 2.
A Ω-δ plot illustrating the phase selection in HEAs ("S" indicates solid solution; "I" indicates intermetallics; "B" indicates amorphous phase) [27].

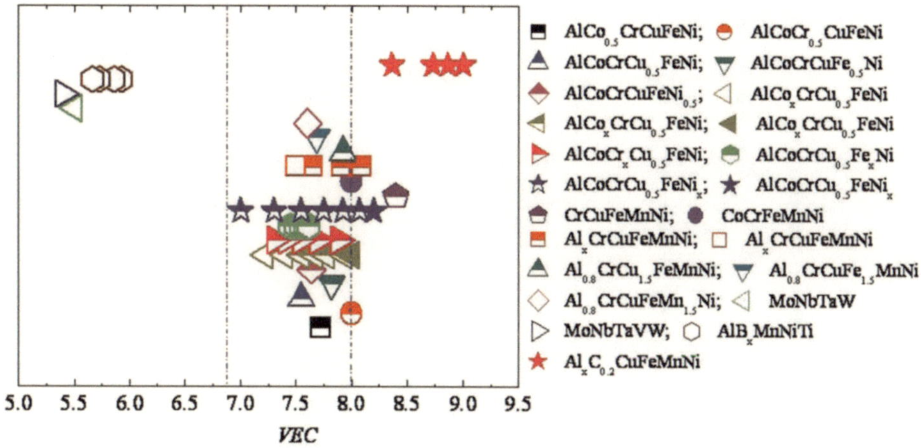

Figure 3.
Relationship between valence electron concentration (VEC) and the stability of FCC and BCC phases in various HEA systems [38].

Guo et al. [31] and Battezzati et al. [37] summarized the relationship between VEC and phase structure in HEAs. As shown in **Figure 3**, HEAs with a single FCC phase typically have VEC values greater than 8.0, while those with a single BCC phase have values below 6.87, and HCP phases occur around a VEC of 2.8.

The aforementioned commonly used parameter rules, along with other physical models, offer effective guidelines for predicting phase types in high-entropy alloys (HEAs). However, it is crucial to note that the overall accuracy of existing physical models remains relatively low, with the highest reported accuracy reaching approximately 72% [24]. Thus, experimental methods are often necessary to validate these predictions.

3. Computational methods

The current computational methods for designing high-entropy alloys (HEAs) primarily include phase diagram calculation (CALPHAD), first-principles calculations (e.g., ab initio and density functional theory (DFT)), molecular dynamics (MD), Monte Carlo simulations (MC). CALPHAD has been extensively utilized in traditional alloy composition design, which involves compiling thermodynamic databases, simulating phase diagrams, predicting phases formed, and employing computational techniques [23]. Thermo-Calc stands out as the predominant software for thermodynamic calculation of HEAs, leveraging CALPHAD-based databases. Recently, Thermo-Calc has successfully contributed to designing high-performance HEAs, such as $(FeCoNi)_{92}Al_{2.5}Ti_{5.5}$ [39] and $CoCrFeNiNb_x$ eutectic HEAs [40], known for superior mechanical properties. For instance, in the case of $CoCrFeNiNb_x$ eutectic HEAs, He et al. proposed that adding Nb to CoCrFeNi forms a pseudo-binary eutectic alloy due to the formation of intermetallics [40]. They developed the (CoCrFeNi)-Nb pseudo-binary diagram using Thermo-Calc, identifying a Nb-10 at.% eutectic composition (**Figure 4**). Accordingly, they designed various $CoCrFeNiNb_x$ ($x = 0.1, 0.25, 0.5, 0.8$) alloys, which exhibit hypo-eutectic, eutectic, and hyper-eutectic structures, respectively. The predictions by Thermo-Calc agreed well with the experimental phase

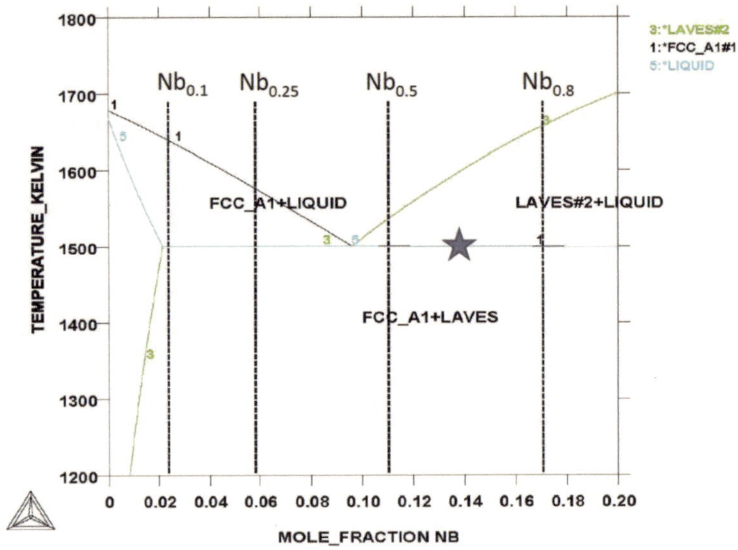

Figure 4.
Phase diagram of the CoCrFeNiNb$_x$ HEAs [40].

diagrams. However, the development of thermodynamic databases using CALPHAD relies heavily on accurate experimental phase diagrams and thermochemical data. Extrapolating from binaries and ternaries to multiple component systems may not always match experimental outcomes, such as the quaternary and/or quinary systems. Given the infancy of HEA research, limited experimental data on phase equilibrium and thermochemistry complicate the direct validation of thermodynamic databases [41].

As another important computational method, the first-principles calculation (e.g. the ab initio calculation using density functional theory (DFT)) augments the CALPHAD method when lacking sufficient thermodynamic data [42–44]. The DFT involves the direct calculation of the electronic structure of atoms through the solution of Schrodinger equations, which can determine formation energies, magnetic states, and lattice parameters for multicomponent phases [23]. However, DFT calculations have uncertainties in handling the d orbitals of transition-metal atoms, which are commonly used in HEAs [45]. Moreover, the DFT calculations require a large amount of numerical computation, which leads to the rapidly increasing computing time with the number of atoms [23]. Other computational methods include molecular dynamics (MD), Monte Carlo (MC), and others. Unlike first-principles calculations, both MD and MC are atomistic techniques. These two methods are mainly used for the prediction of structures of initial cluster formation or the material response to stimuli rather than phase prediction [23].

In recent years, machine learning (ML) has been increasingly employed to accelerate the simulation and phase prediction of HEAs [45–50]. On the one hand, utilizing DFT datasets, machine learning (ML) methods can construct atomic interaction models (AIMs) that effectively describe complex interactions within multicomponent alloys. This approach circumvents the need for time-consuming first-principles thermodynamics simulations. This method eliminates the need for lengthy first-principles thermodynamics simulations. Additionally, machine learning can be utilized to develop models for predicting bulk properties, including phase constituents, crystal structures, yield strengths, etc. These models are built using high-throughput experimental or computational data, demonstrating the versatility of ML in materials science research [51]. For instance, Zheng et al. applied ML methods to design nanoprecipitate-strengthened HEAs using extensive thermodynamic data [46]. The newly developed γ'-strengthened HEA demonstrates enhanced mechanical properties, exhibiting a yield strength of approximately 1.31 GPa and a tensile elongation of about 15%. In addition, Su et al. introduced an optimized ML model that achieved high accuracy rates, reaching up to 88.7% in identifying solid-solution and non-solid-solution HEAs, and further achieving 91.3% accuracy in distinguishing BCC, FCC, and dual-phase HEAs [47]. These studies present an alternative approach to HEA design, accelerating the development of novel HEAs with superior performance.

4. Self-optimizing methods

In contrast to traditional alloys based on a single principal element, the multicomponent approach of HEAs offers an almost limitless range of alloy systems. Experimental testing of all these alloys is impractical. To date, just over 400 HEA systems have been reported, primarily developed through trial-and-error experiments and modeling methods [23]. Consequently, there is a pressing need to devise new strategies that can efficiently identify potential high-performance alloy compositions from the vast HEA compositional space. Tracing back to the origins of the renowned CoCrFeMnNi HEA, Brian Cantor, one of the pioneers in HEA research, has applied a rapid alloy design strategy. This approach, later termed the "self-optimizing strategy" by Yin and co-workers [25], involves identifying the chemical composition of supersaturated single solid-solution phases by directly using the local compositions of specific phase constituents within existing multicomponent alloys containing dual or multiple phases. Brian Cantor et al. initially developed a 20-component alloy with an equiatomic chemical composition of MgAlSiMnCrFeCoNiCuZnGeNbMoAgCdSnSbWPbBi, which exhibited a multi-phase structure (**Figure 5a**). However, one of the phases in this alloy predominantly consisted of five components in nearly equal proportions (i.e. Cr, Mn, Fe, Co, and Ni), confirmed to be a single FCC phase upon producing a new sample using the probed composition. Therefore, while the equiatomic composition of CoCrFeMnNi was not deliberately designed by Cantor

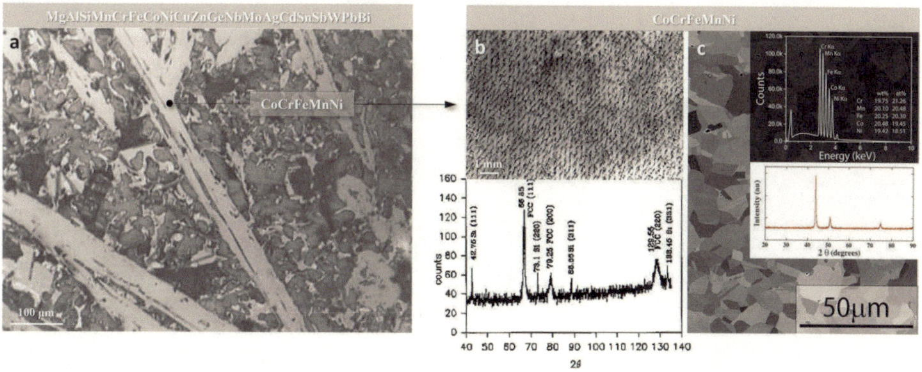

Figure 5.
a. Optical micrograph of the equiatomic 20-component alloy MgAlSiMnCrFeCoNiCuZnGeNbMoAgCdSnSbWPbBi [52]; b. Microstructure and XRD patterns of the as-cast CrMnFeCoNi alloy [2]; c. SEM image, EDS analysis, and XRD patterns of the fully recrystallized CrMnFeCoNi alloy [53].

et al., it emerged naturally from the alloy itself through the inherent solute partitioning or diffusion processes among its constituent elements [2]. As shown in **Figure 5b**, the as-cast Cantor alloy, which has a single FCC solid-solution phase, exhibits a typical dendritic microstructure with Cr, Fe, and Co segregation along the dendrites. Thermomechanical processes, such as annealing after cold rolling, can transform the dendritic structure of the as-cast Cantor alloy into a homogeneous microstructure with equiaxed grains and annealing twins (**Figure 5c**). Cantor et al. not only ushered in the research era of high-entropy alloys but also pioneered a rapid alloy design strategy. However, since the advent of CoCrFeMnNi Cantor alloy, researchers appear to have been solely focused on the novel concept of HEAs, overlooking this innovative rapid alloy design strategy.

In 2024, inspired by the intermediate-temperature instability of HEAs, Yin et al. proposed a novel approach for rapidly designing Nanostructured Multicomponent (NM) alloys [25]. Using the FeAlTiVCrMoW HEA as a model, Yin et al. firstly destabilized the BCC matrix into two phases by increasing the Fe content (**Figure 6b-d**). This led to the development of a new Fe5AlTiVCrMoW HEA with dual

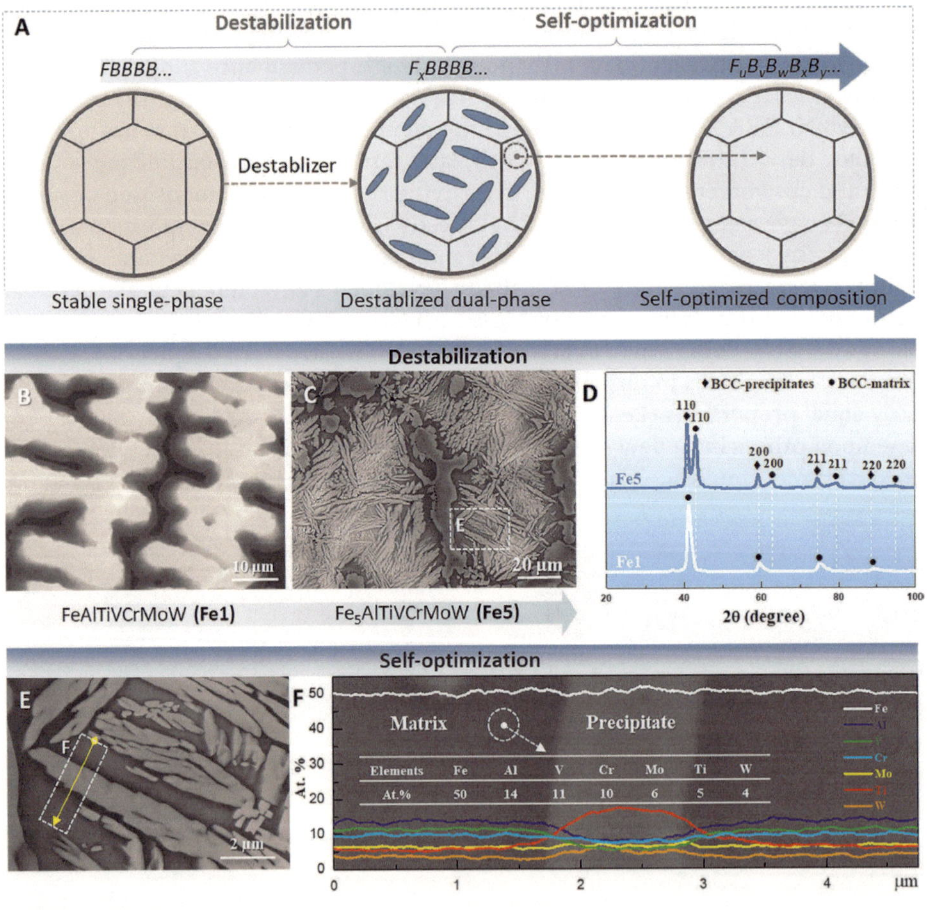

Figure 6.
A. Schematic illustration of the alloy design strategy; B. Microstructure of the as-cast FeAlTiVCrMoW HEA; C. Microstructure of the as-cast Fe_5AlTiVCrMoW HEA. D. XRD patterns of the FeAlTiVCrMoW; E. Magnified SEM image of the as-cast Fe_5AlTiVCrMoW HEA; F. EDS spectra of the BCC matrix and precipitates in Fe_5AlTiVCrMoW HEA [25].

BCC phases. Next, the "self-optimizing" method was applied by directly leveraging the local composition of the metastable phase within the destabilized dual-phase Fe5AlTiVCrMoW HEA (**Figure 6e-f**). The novel self-optimized alloy with a composition of Fe50Al14V11Cr10Mo6Ti5W4 exhibited a two-stage phase decomposition behaviour at 600°C, showcasing a strong ageing response. The peak-aged hardness can be up to ~1378 HV. This was attributed to the formation of a distinctive multiphasic nanostructure, consisting of two BCC phases and one B2 phase, which significantly increased the density of phase boundaries and enhanced the strengthening effect [25]. This approach enables the effective design of large-scale NM alloys based on HEAs, which can be synthesized using conventional techniques such as casting and heat treatment. In contrast to the conventional focus on attaining a single solid-solution phase in HEAs, this study introduces an alternative method for designing bulk nanostructured alloys with complex compositions and multiple phases, thereby achieving superior properties.

5. Conclusions and outlook

In recent decades, several strategies have emerged to explore the vast composition space of HEAs and identify compositions that result in superior properties. These include physical modeling, computational approaches, and self-optimizing strategies. Among the physical modeling methods, parameter rules such as the ΔH_{mix} and δ rules, the Ω and δ rule, and the VEC rule have been widely utilized due to their simplicity and relatively high accuracy in predicting solid solution phases. These rules have proven effective in composition design and phase prediction. However, the overall accuracy of existing physical models remains relatively limited, particularly in predicting the phase constituents of precipitates.

In contrast, computational methods offer potentially higher accuracy in predicting phases of HEAs. Among these methods, CALPHAD are extensively employed. However, their accuracy diminishes as alloy complexity increases beyond binaries and ternaries, owing to limited experimental phase diagrams and thermochemical data. First-principles calculations, such as DFT, complement CALPHAD when thermodynamic data is insufficient. Nevertheless, DFT calculations introduce uncertainties, particularly in handling d orbitals of transition-metal atoms prevalent in HEAs. Moreover, the DFT calculations involve intensive numerical computations, with computational demands escalating with system size. To accelerate simulations and enhance phase prediction accuracy for HEAs, machine learning (ML) has gained prominence. ML sidesteps the need for time-consuming first-principles simulations, directly predicting bulk properties such as phase formations, crystal structures, and mechanical properties.

In addition to traditional physical modeling and computational methods, a novel composition design concept has emerged: the "self-optimizing strategy". Inspired by the discovery of the Cantor alloy, this approach involves identifying the chemical composition of supersaturated single solid-solution phases by directly utilizing the local compositions of specific phase constituents within existing multicomponent alloys containing dual or multiple phases. This "self-optimizing" alloy design strategy circumvents both experimental and computational "trial-and-error" processes. It can be applied to conventional or multicomponent alloys with dual or multiple phases, facilitating the rapid design or optimization of compositions, customized microstructures, and improved mechanical properties.

Acknowledgements

This work is financially supported by the Australian Research Council Discovery Project DP200101408 (MXZ) and Sun Yat-sen University.

Conflict of interest

The authors declare no conflict of interest.

Author details

Yu Yin[1,2], Libo Zhou[1], Dekui Mu[3], Han Huang[1*], Mingxing Zhang[2*] and Huapan Xiao[1]

1 School of Advanced Manufacturing, Sun Yat-sen University, Shenzhen, China

2 School of Mechanical and Mining Engineering, University of Queensland, Brisbane, Australia

3 Institute of Manufacturing Engineering, Huaqiao University, Xiamen, China

*Address all correspondence to: hanhuang@sysu.edu.cn; mingxing.zhang@uq.edu.au

References

[1] Yeh JW et al. Nanostructured high-entropy alloys with multiple principal elements: Novel alloy design concepts and outcomes. Advanced Engineering Materials. 2004;**6**(5):299-303

[2] Cantor B et al. Microstructural development in equiatomic multicomponent alloys. Materials Science and Engineering a-Structural Materials Properties Microstructure and Processing. 2004;**375**:213-218

[3] Huang PK et al. Multi-principal-element alloys with improved oxidation and wear resistance for thermal spray coating. Advanced Engineering Materials. 2004;**6**(12):74-78

[4] Zhang Y et al. Microstructures and properties of high-entropy alloys. Progress in Materials Science. 2014;**61**: 1-93

[5] Ye YF et al. High-entropy alloy: Challenges and prospects. Materials Today. 2016;**19**(6):349-362

[6] Miracle DB, Senkov ON. A critical review of high entropy alloys and related concepts. Acta Materialia. 2017;**122**: 448-511

[7] Chen J et al. A review on fundamental of high entropy alloys with promising high–temperature properties. Journal of Alloys and Compounds. 2018;**760**: 15-30

[8] George EP, Raabe D, Ritchie RO. High-entropy alloys. Nature Reviews Materials. 2019;**4**(8):515-534

[9] George EP, Curtin WA, Tasan CC. High entropy alloys: A focused review of mechanical properties and deformation mechanisms. Acta Materialia. 2020;**188**: 435-474

[10] Yeh J-W. Recent progress in high-entropy alloys. Annales de Chimie Science des Matériaux. 2006;**31**(6): 633-648

[11] Li Z et al. Metastable high-entropy dual-phase alloys overcome the strength-ductility trade-off. Nature. 2016; **534**(7606):227-230

[12] Lu W, Liebscher CH, Dehm G, Raabe D, Li Z. Bidirectional transformation enables hierarchical nanolaminate dual-phase high-entropy alloys. Advanced Materials. 2018;**30**(44): 1804727

[13] Li ZM et al. A TRIP-assisted dual-phase high-entropy alloy: Grain size and phase fraction effects on deformation behavior. Acta Materialia. 2017;**131**: 323-335

[14] Li JS et al. Enhanced mechanical properties of a CoCrFeNi high entropy alloy by supercooling method. Materials & Design. 2016;**95**:183-187

[15] Lei Z, Liu X, Wu Y, Wang H, Jiang S, Wang S, et al. Enhanced strength and ductility in a high-entropy alloy via ordered oxygen complexes. Nature. 2018;**563**(7732):546-550

[16] Zou Y, Ma H, Spolenak R. Ultrastrong ductile and stable high-entropy alloys at small scales. Nature Communications. 2015;**6**: 7748

[17] Huang H, Wu Y, He J, Wang H, Liu X, An K, et al. Phase-transformation ductilization of brittle high-entropy alloys via metastability engineering. Advanced Materials. 2017;**29**(30): 1701678

[18] He JY et al. A precipitation-hardened high-entropy alloy with outstanding

tensile properties. Acta Materialia. 2016; **102**:187-196

[19] Wani IS et al. Ultrafine-grained AlCoCrFeNi2.1Eutectic high-entropy alloy. Materials Research Letters. 2016; **4**(3):174-179

[20] Yin Y et al. Novel cost-effective Fe-based high entropy alloys with balanced strength and ductility. Materials & Design. 2019;**162**:24-33

[21] Yin Y, Ren W, Tan Q, Chen H, Huang H, Zhang MX. A cost-effective cryogenic high-entropy alloy with high strength-ductility synergy and strain hardenability. Materials Science and Engineering: A. 2023;**865**:144607

[22] Fan L et al. Ultrahigh strength and ductility in newly developed materials with coherent nanolamellar architectures. Nature Communications. 2020;**11**(1):6240

[23] Murty BS et al. High-Entropy Alloys. Amsterdam: Elsevier; 2019

[24] Li J-H, Tsai M-H. Theories for predicting simple solid solution high-entropy alloys: Classification, accuracy, and important factors impacting accuracy. Scripta Materialia. 2020;**188**: 80-87

[25] Yin Y et al. A new route to bulk nanostructured multiphase alloys with ultrahigh hardness. Journal of Materials Science & Technology. 2025;**210**:151-158

[26] Wei S, Kim SJ, Kang J, Zhang Y, Zhang Y, Furuhara T, et al. Natural-mixing guided design of refractory high-entropy alloys with as-cast tensile ductility. Nature Materials. 2020;**19**(11): 1175-1181

[27] Yang X, Zhang Y. Prediction of high-entropy stabilized solid-solution in multi-component alloys. Materials Chemistry and Physics. 2012;**132**(2–3): 233-238

[28] Pickering EJ, Jones NG. High-entropy alloys: A critical assessment of their founding principles and future prospects. International Materials Reviews. 2016;**61**(3):183-202

[29] Tsai MH, Yeh JW. High-entropy alloys: A critical review. Materials Research Letters. 2014;**2**(3): 107-123

[30] Alaneme KK, Bodunrin MO, Oke SR. Processing, alloy composition and phase transition effect on the mechanical and corrosion properties of high entropy alloys: A review. Journal of Materials Research and Technology-Jmr&T. 2016; **5**(4):384-393

[31] Guo S, Liu CT. Phase stability in high entropy alloys: Formation of solid-solution phase or amorphous phase. Progress in Natural Science: Materials International. 2011;**21**(6):433-446

[32] Takeuchi A, Inoue A. Quantitative evaluation of critical cooling rate for metallic glasses. Materials Science and Engineering a-Structural Materials Properties Microstructure and Processing. 2001;**304**(Supplement C): 446-451

[33] Takeuchi A, Inoue A. Classification of bulk metallic glasses by atomic size difference, heat of mixing and period of constituent elements and its application to characterization of the main alloying element. Materials Transactions. 2005; **46**(12):2817-2829

[34] Guo S et al. More than entropy in high-entropy alloys: Forming solid solutions or amorphous phase. Intermetallics. 2013;**41**:96-103

[35] Zhang Y et al. Solid-solution phase formation rules for multi-component alloys. Advanced Engineering Materials. 2008;**10**(6):534-538

[36] Kittel C, McEuen P, McEuen P. Introduction to Solid State Physics. Vol. 8. New York: Wiley; 1996

[37] Poletti MG, Battezzati L. Electronic and thermodynamic criteria for the occurrence of high entropy alloys in metallic systems. Acta Materialia. 2014; **75**(Supplement C):297-306

[38] Guo S et al. Effect of valence electron concentration on stability of fcc or bcc phase in high entropy alloys. Journal of Applied Physics. 2011; **109**(10):103505

[39] Guo L et al. CALPHAD aided design of high entropy alloy to achieve high strength via precipitate strengthening. Science China Materials. 2020;**63**(2): 288-299

[40] He F et al. Designing eutectic high entropy alloys of CoCrFeNiNbX. Journal of Alloys and Compounds. 2016;**656** (Supplement C):284-289

[41] Zhang C, Gao MC. CALPHAD modeling of high-entropy alloys. In: Gao MC et al., editors. High-Entropy Alloys: Fundamentals and Applications. Cham: Springer International Publishing; 2016. pp. 399-444

[42] Ikeda Y, Grabowski B, Körmann F. Ab initio phase stabilities and mechanical properties of multicomponent alloys: A comprehensive review for high entropy alloys and compositionally complex alloys. Materials Characterization. 2019; **147**:464-511

[43] Li X et al. Ab initio-predicted micro-mechanical performance of refractory high-entropy alloys. Scientific Reports. 2015;**5**:12334

[44] Li ZM et al. Ab initio assisted design of quinary dual-phase high-entropy alloys with transformation-induced plasticity. Acta Materialia. 2017;**136**: 262-270

[45] Huang W, Martin P, Zhuang HL. Machine-learning phase prediction of high-entropy alloys. Acta Materialia. 2019;**169**:225-236

[46] Zheng T, Hu X, He F, Wu Q, Han B, Chen D, et al. Tailoring nanoprecipitates for ultra-strong high-entropy alloys via machine learning and prestrain aging. Journal of Materials Science & Technology. 2021;**69**:156-167

[47] Zhang Y et al. Phase prediction in high entropy alloys with a rational selection of materials descriptors and machine learning models. Acta Materialia. 2020;**185**:528-539

[48] Wu Q et al. Uncovering the eutectics design by machine learning in the Al–Co–Cr–Fe–Ni high entropy system. Acta Materialia. 2020;**182**:278-286

[49] Wen C et al. Machine learning assisted design of high entropy alloys with desired property. Acta Materialia. 2019;**170**:109-117

[50] Kim G et al. First-principles and machine learning predictions of elasticity in severely lattice-distorted high-entropy alloys with experimental validation. Acta Materialia. 2019;**181**: 124-138

[51] Liu X, Zhang J, Pei Z. Machine learning for high-entropy alloys: Progress, challenges and opportunities. Progress in Materials Science. 2023;**131**: 101018

[52] Vincent A. A Study of Three Multicomponent Alloys (BSc Part II Thesis). UK: University of Sussex; 1981

[53] Gludovatz B et al. A fracture-resistant high-entropy alloy for cryogenic applications. Science. 2014; **345**(6201):1153-1158

Chapter 2

High Entropy Alloy Composition Design for Mechanical Properties

Amir H. Naghdi, Dario Massa, Kamran Karimi and Stefanos Papanikolaou

Abstract

Multi-component high-entropy alloys (HEAs) are a novel class of materials exhibiting outstanding material properties that often surpassing their traditional counterparts. Despite their ubiquity, the underlying microstructure-property relationships in HEAs remain elusive. This chapter addresses this gap by exploring the application of cutting-edge machine learning tools to establish robust connections between HEAs' chemical composition, microstructure, and mechanical response. The survey begins by discussing the current state of micro-structural characterization techniques in HEAs, giving insights into their complex underlying microstructure. The development of ML force fields for HEAs based on ab initio datasets is then highlighted, addressing challenges posed by the expansive composition space associated with HEAs. The chapter further outlines machine learning-assisted composition search strategies for HEAs with specific functional properties, offering a systematic and efficient approach to explore material properties. Overall, the present overview demonstrates the potential of machine learning in unraveling the intricate nature of HEAs and accelerating their tailored design for diverse applications.

Keywords: high-entropy alloys, thermo-mechanical properties, micro-structural characterization, machine learning applications, composition design

1. Introduction

Multi-component high-entropy alloys (HEAs) represent an emerging class of materials that are composed of at least four or more elements in roughly equal proportions. As opposed to traditional alloys, typically consist of one or two primary elements with smaller amounts of additional elements, HEAs exhibit exceptional thermo-mechanical properties that often outperform the former alloys [1–3]. HEAs' transport properties, for instance, have long been characterized by comparatively slow kinetics, as opposed to pure metals and conventional alloys, hence the term sluggish diffusion [4, 5]. Together with high entropy of mixing, severe lattice distortion, and also, the cocktail effect, these so-called "core effects" are commonly identified as the principal sources of HEAs' superior performance

 IntechOpen

(*e.g.* single-phase thermodynamic stability [6], creep resistance [7], and high-temperature strength [8]).

The atomistic origin of these remarkable properties presumably originate from the underlying chemical/micro-structural complexities attributed to HEAs, yet the nature of these inherent microstructure-property relationships remain elusive to this date. Empirical frameworks conventionally describe such correlations by a fairly small set of "descriptors" largely ignoring underlying scale hierarchies and intricate micro-structure. Multi-scale simulation frameworks have fairly limited applicability/predictability due to the modeling gap in transferring physics-based information across different length/time-scales. Experimental investigations can only explore a small portion of the immense combinatorial space spanned by different elemental compositions in HEAs.

The above limitations demand applications of state-of-the-art machine learning (ML) tools that can help establish robust relationships between HEAs' composition and their mechanical response. This chapter aims to conduct a thorough survey of the current state-of-the-art in data mining and pattern detection, feature extraction and analysis, physics extraction and interpretation of ML predictions relevant to HEAs' composition design and properties. A special emphasis will be given to applications of machine learning relevant to characterization of micro-structural defects, development of machine-learned interatomic potentials via ab initio calculations, and ml-assisted composition search strategies for targeted functional properties.

The organization of this chapter is as follows. Section 2 discusses the current state of micro-structural characterization for HEAs by employing advanced imaging techniques such as digital image correlation, X-ray diffraction, and high-resolution microscopy. Material scientists utilize these techniques, along with sophisticated image processing tools and machine learning-inspired descriptors, to extract meaningful insights into structural patterns, physics, and property correlations from complex micro-structural data in HEAs. Section 3 highlights the development process of machine learning force fields for HEAs based on ab initio datasets. We present relevant challenges arising from the expansive composition space associated with HEAs and certain mitigation strategies. In Section 4, we review machine learning-assisted composition search strategies for HEAs with specific functional properties, addressing challenges in navigating their vast compositional space. Our focus will be on the efficacy and robustness of the machine learning-supported HEA design, as opposed to traditional methods like density functional theory and CALPHAD, offering a systematic and efficient approach to explore material properties and guiding experimental validation.

2. Data mining, pattern detection, and physics extraction in HEAs

Current state of the art in the high-throughput design of HEAs involves extensive applications of powerful characterization techniques which allow for gaining deep insights into inherent structural patterns, underlying physics, and structure-property correlations across a broad range of length scales [9]. Material scientists make routine use of advanced image-based techniques, such as digital image correlation (DIC), X-ray diffraction (XRD) scattering, and high-resolution scanning/transmission microscopy (STEM), capturing two-dimensional micro-structural information at micro and nano scales down to atomic resolutions. These data sets, along with simulation-generated data, often embed complex hierarchical patterns and necessitate the employment of advanced image processing tools, *i.e.* for tasks related to the analysis of complex atomic displacement patterns [10, 11], direct recognition of plasticity through local strain evolution [12–15] as well as identification/classification of local chemical environments and structural ordering [16–20]. Extracting meaningful insights from micro-structural data through data analytic and feature-extraction algorithms is vitally important for HEAs. The present section highlights certain effective tools, some of them inspired by machine learning, that can help address the physics extraction aspect of HEAs.

2.1 General structural descriptors

As previously outlined, the aforementioned characterization techniques offer hierarchical insights into the underlying materials' substructure and, therefore, the notion of relevant structural descriptors is highly scale-dependent [22]. Atomically-resolved images, obtained from high-resolution TEM typically with Å-level resolution, allow for extracting individual atomic positions $\vec{r}_i$, $i = 1, 2, \ldots$ with remarkable accuracy. Such high-precision coordinates may be subsequently utilized to construct atom-based metrics and/or spatially-averaged proxies of the underlying microstructure. In this context, one may analyze statistical order parameters including two-point pair-correlation function in real space $g\left(\vec{r}\right)$ and/or structure factor $S\left(\vec{q}\right)$ [23], exploiting the overall periodicity within the crystalline structure, and seek for underlying contrasts within the local structure. The latter relates to diffraction pattern intensities $D\left(\vec{q}\right) \propto \left|S\left(\vec{q}\right)\right|^2$ as the key metric to describe crystal symmetries [24]. This methodology is presented in **Figure 1** illustrating a schematic two-dimensional diffraction pattern and the resulting contrast owing to the defective structure [21].

Direct recognition of these patterns via machine learning tools has proven to be useful for image-based microstructure classification tasks in materials science [25–27] as well as crystallography and phase segmentation [28], and material property predictions [29–31]. Along these lines, a machine learning-based approach has been developed by Kaufmann *et al.* for phase selection demonstrating a rapid and autonomous method for identifying crystal symmetry based on diffraction patterns [27, 32]. Within the context of HEAs, establishing such relationships has proven challenging

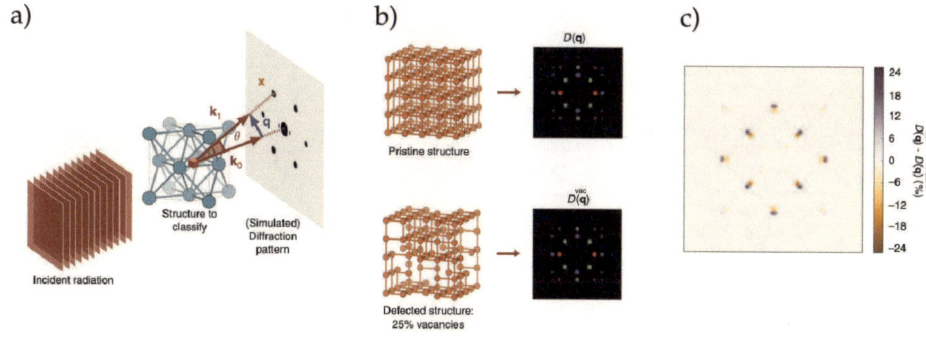

Figure 1.
Two-dimensional diffraction pattern is schematically represented in a). Panel b) illustrates a pristine simple cubic structure and the same structure with vacancies along with corresponding diffraction fingerprints. The difference between the diffraction fingerprints of the defective structure and the pristine structure is depicted in c). Figure taken from [21].

due to the high configurational entropy, leading to a distribution of lattice parameters and cell compositions instead of a singular unit cell and lattice constant typically observed in traditional alloys [33]. In [34], the authors proposed a topological approach analyzing experimental atom probe tomography (APT) data and atomic neighborhoods in Al1.3CoCrCuFeNi and Al0.3CoCrFeNi HEAs. Despite the presence of severe lattice distortion, their methodology achieved a high accuracy in classifying the lattice structure of atomic neighborhoods including BCC and FCC phases.

2.2 Local descriptors for chemical short-range ordering

To investigate the local chemical environment (including short-range chemical ordering), the Warren–Cowley SRO parameters $\alpha_{ab}\left(\vec{r}\right)$ [35] provide robust chemical/structural metrics probing concentration variations of type-b atoms within distance $\vec{r}$ from a center type-a element. This metric is applicable to experimentally derived datasets (i.e. like atom probe tomography) [36–38] and/or simulation-based datasets [19, 39–41], containing both atomic identities and three-dimensional coordinates. In this framework, any coherent compositional deviations from (statistically) random distributions of atoms within the solution matrix, the so-called random solid-solution limit, can be regarded as spatial footprints of chemical ordering [16, 37, 42, 43]. Theses order parameters relate to the spatial density correlations with the associated two-point correlation function defined as $c_{ab}\left(\vec{r}-\vec{r}'\right)=\left\langle\rho_a\left(\vec{r}\right)\rho_b\left(\vec{r}'\right)\right\rangle$. Here the angle brackets $\langle . \rangle$ denote ensemble averaging and $\rho_a\left(\vec{r}\right)$ refers to the local concentration of type-a element. **Figure 2** presents direct evidence of angstrom-level chemical short-range order in an annealed NiCoV medium-entropy alloy and associated correlation analysis featuring meaningful (anti-)correlation patterns in space.

The SRO patterns primarily arises from processing parameters (e.g., special heat treatment, alloy composition), yet mechanical deformation on pre-annealed alloys was further reported to facilitate the SRO formation [44]. Notably, Naghdi *et al.* [45] employed molecular dynamics simulations to demonstrate that nano-indentation, specifically dwell nano-indentation protocols, can manipulate chemical short-range

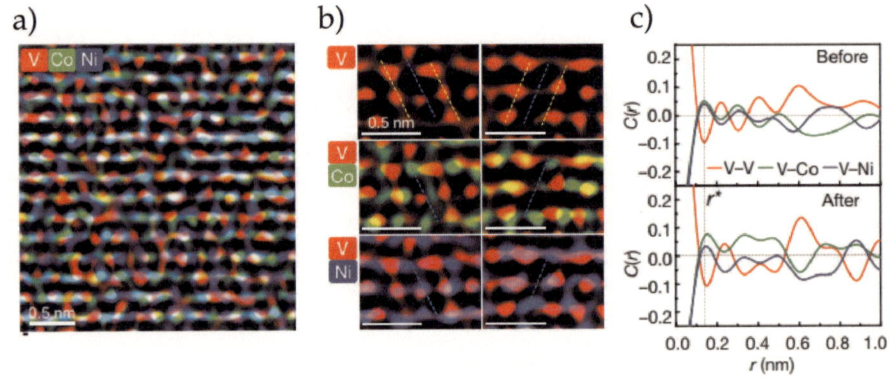

Figure 2.
Chemical mapping in (a) illustrating the local distribution of elements in annealed face-centered-cubic NiCoV concentrated solution. Panel (b) provides magnified views of V, V-Co, and V-Ni elements. The corresponding pair correlation coefficients, $c_{ab}(r)$, are presented in (c) before and after tensile tests, offering insights into SRO distributions in space. Figure reproduced from [16].

order in equi-atomic NiCoCr alloys. Their research revealed the formation of density-wave stripe patterns under the indenter tip, mainly influenced by local stress concentrations, providing explicit validation pathways for the manipulation of SROs in multicomponent alloys. In [40, 46], combined lattice distortion and SRO effects were reported to enhance the critical resolved shear stress associated with simulated NiCoV and NiCoCr concentrated solid solution alloys.

In studying NiCoCr-based complex alloys, stacking fault energy, hardness, and fracture toughness, as bulk properties, were recently shown to strongly correlate with the degree of Ni-rich SROs and corresponding structural features [39, 41, 42, 47]. On the other hand, local misfit properties and associated atomic-scale fluctuations have been long known to serve as favored trapping sites for dislocation migration with appreciable contributions to plastic yielding properties [48–50]. Current efforts for direct measurements of local lattice misfit volumes and deviatoric strains in CSAs have been mostly limited to atomistic simulations. In this context, direct experimental characterization of local distortion properties and SROs and their collective effects on the dislocation mobility in HEAs is still largely unexplored.

2.3 Insights from microscopic deformation in polycrystalline alloys

Electron backscatter diffraction (EBSD) imaging provides detailed local crystallography information at each specific location on the studied surface and enables a thorough examination of micro texture and sample morphology down to sub-micron levels [51, 52]. These maps delineate distinct crystallographic signatures, aiding in the phase identification [32], characterization of local strain and deformation patterns [53] as well as twinning [54], (geometrically necessary) dislocations [55], and other crystalline defects within grains substructure. EBSD datasets typically contain pixel-based orientation information with a relatively high precision (order 0.1°) and fine spatial resolution (order 100 nm) which can be conveniently processed by various statistical tools and machine learning algorithms (**Figure 3**).

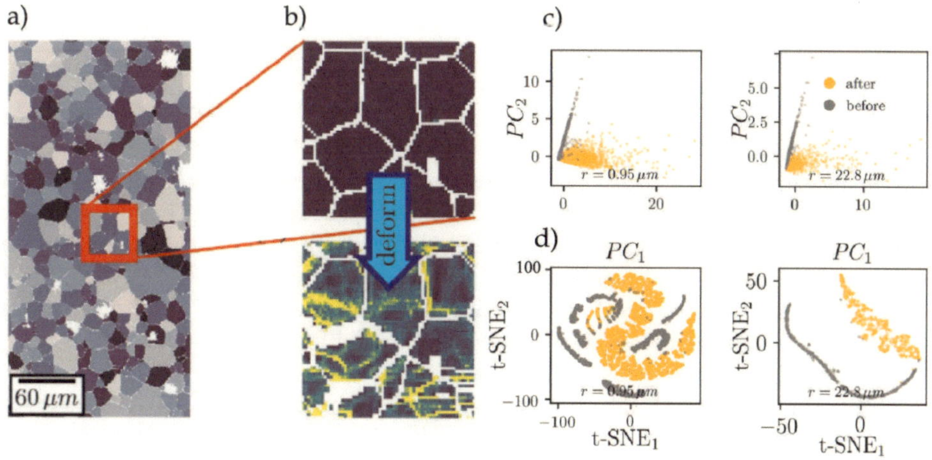

Figure 3.
EBSD imaging serves as a foundation for machine learning applications. Panel (a) presents a snapshot of the grain structure for Mg alloy. In (b), the sample was characterized by the dislocation density (ρ_{GND}) both before and after deformation. The panels in (c) and (d) serve as representatives for the machine learning dataset, encompassing the two classes associated with grains before and after deformation. Figure taken from [56].

In conjunction with nano/micro-mechanical deformation tests, such datasets can give further insights into the intricate links between complex surface characteristics and mechanical behavior at microscopic levels [57, 58]. In this framework, Salmenjoki *et al.* [56] examined the evolution of dislocation density in polycrystalline magnesium-zinc (Mg-Zn) alloys by comparing EBSD images before and after deformation. The study aimed to predict the dislocation density levels upon deformation solely based on associated changes in grain microstructure. Based on the grain-wise information prior to loading, machine learning Graph Neural Network (GNN) models were trained representing the grain microstructure to predict geometrically necessary dislocation density within the deformed alloy. Utilizing a GNN model, a similar approach was employed in [59] investing predictability of nanoscale hardness in polycrystalline metals based on micro-structural features such as grain orientations and neighboring grain properties.

2.4 Machine learning descriptors for diverse microstructures

Machine learning-based descriptors play a pivotal role in the context of computational chemistry and materials science, enabling an efficient (yet accurate) representation of complex molecular and material microstructure [60, 61]. Among the notable descriptors, the Behler-Parinello method [62] and relevant atom-centered symmetry functions (ACSFs) stand out for their capability to capture the inherent symmetry and non-local features in atomic environments. The radial ACSF represents a two-body function that can be constructed using a Gaussian-smoothed atomic density function $\rho_i(r) = \sum_j \exp\{-(r_{ij} - r)/2\sigma^2\}$ centered at atom i. Here r_{ij} denotes the pair-wise distance between center atom i and a neighbor atom j and σ is the width of the Gaussian kernel. The angular ACSF include more tunable parameters leading to an immense flexibility of the function. This includes three-body terms corresponding to the angle between the vectors $\vec{r}_{ij}$ and $\vec{r}_{ik}$ with j and k being neighboring atoms of center atom i. The Smooth Overlap of Atomic Positions (SOAP) descriptor, on the other hand, excels in representing local atomic environments by making an expansion of the local atomic density based on a complete set of the spherical harmonics $Y_{lm}(\phi, \theta)$ and radial basis functions $g_n(r)$.

In terms of ML-based characterization of defects, a fairly universal strategy for crystalline solids at the atomic scale is to seek for statistical "outliers" within the multi-dimensional descriptor space. This approach is illustrated in **Figure 4** providing a generic measure to describe the *distortion* score of local atomic environments [63]—a metric quantifying an outlier's distance from the observed average. The proposed strategy enables automatic detection of defects, facilitating a hierarchical description of defects [64] and offering applications in advanced materials modeling, including relevant information selection for evaluating energy barriers [65–67], design of robust interatomic machine learning potentials [68–70], and high-throughput analysis of databases for materials science advancements.

Within the context of HEAs, one promising avenue pertains to diffusion studies of micro-structural defects, typically exhibiting sluggish dynamics under extreme conditions (i.e. irradiation effects) [71]. A recent kinetic Monte Carlo (kMC) study probing the micro-structural kinetics of NiCoCr and NiCoCrFeMn alloys revealed a subdiffusive dynamics of atomic vacancies [72]. This subdiffusive phenomenon was interpreted as a universal signature of dynamical sluggishness in HEAs rooted in severe lattice distortion and chemical complexities. The concept of a distortion score is

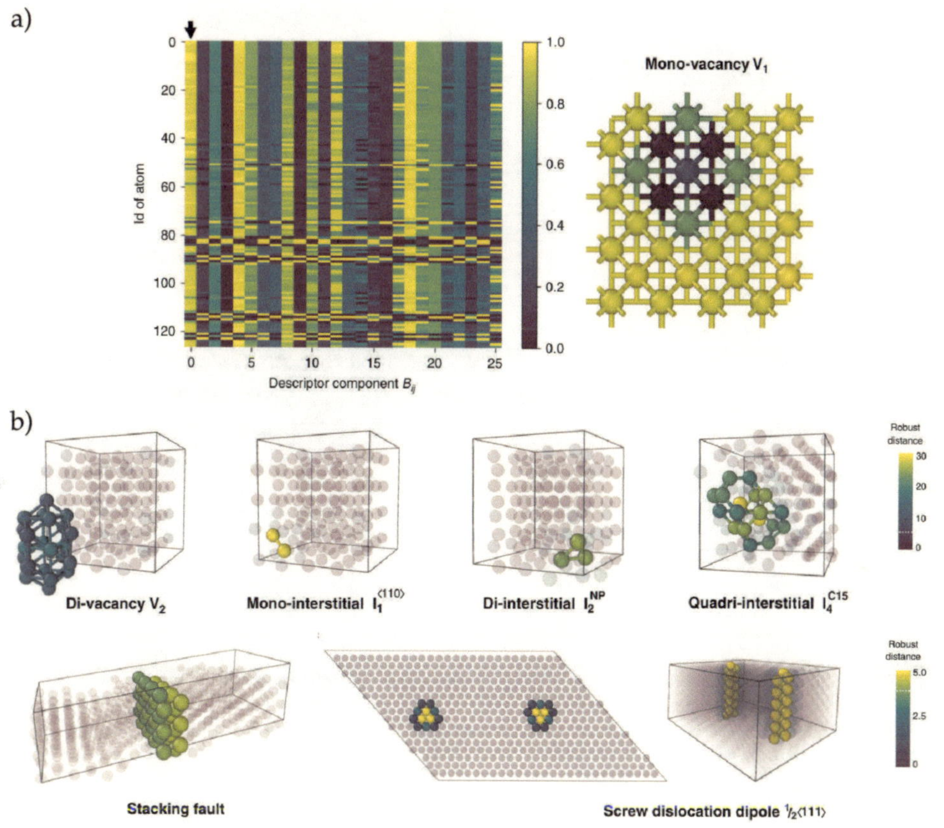

Figure 4.
A single vacancy in a body-centered cubic metal is depicted in the 26-dimensional descriptor space in (a) utilizing a variant of the SOAP descriptor. The spectral map illustrates contrasting bands arising from atomic environments in the vicinity of the mono-vacancy. Here the atoms are colored based on the first descriptor component indicated by the arrow. Panel (b) illustrates several defects in body-centered cubic crystal which are quantified based on the associated distortion scores. Figure taken from [63].

amenable to the automated classification of (irradiation-induced) defects in atomistic configurations. Furthermore, it facilitates the prediction of associated diffusion paths and energy barriers under thermal activation, which is otherwise a formidable task owing to inherent lattice disorder.

3. Machine-learned force fields

3.1 Overview

Machine Learning Force Fields (MLFFs) have emerged as a viable alternative to empirical potentials, such as the Embedded Atom Model (EAM) and Modified Embedded Atom Model (MEAM), demonstrating the capability to achieve quantum accuracy in predicting forces and energies during large length- and time-scale molecular dynamics (MD) simulations [73–87]. Various methods have been developed to date, utilizing diverse machine learning algorithms for MLFF development. These approaches include Gaussian Approximation Potentials (GAP) [88], Neural Network

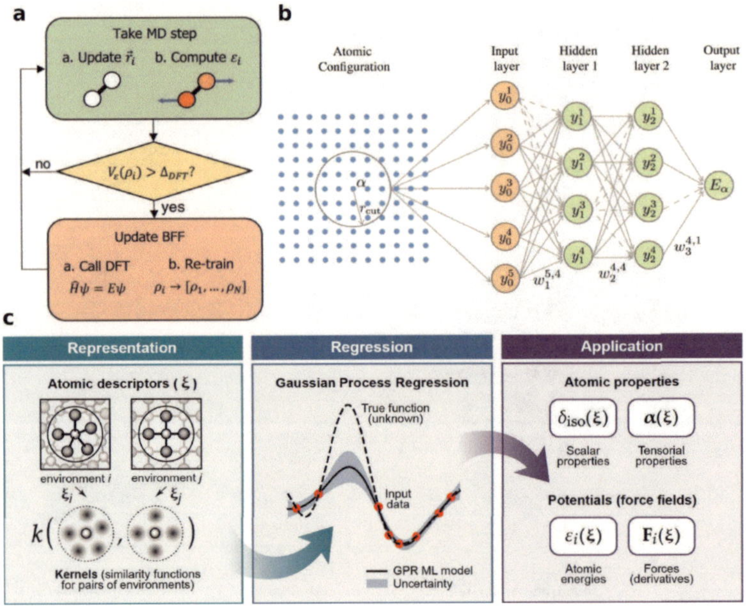

Figure 5.
MLFF overview. (a) Demonstrates an active learning framework for MLFF developement, where new congifurations are added to the training dataset based on intrinsic uncertainty values of standard gaussian process (SGP) model (figuare taken from [81]). (b) Illustrates a NNIP developement workflow, where descriptors are calculated for each atom and are fed to the input leyer of the model. The properties, such as energies and forces, are then predicted from the output layer (figure from [91]). (c) The GAP potential framework, which uses a kernel function to evaluate uncertainties of the model (figure from [92]).

Interatomic Potentials (NNIPs) [82], Tabulated GAP (tabGAP) [89, 90], on-the-fly active learning of interatomic potentials such as FLARE [81] and equivariant graph neural network potentials (NequIP) [80].

MLFF development involves constructing a dataset derived from Ab initio Molecular Dynamics (AIMD) simulations, followed by training a machine learning (ML) algorithm to predict energies and forces for individual atoms within the training dataset (see **Figure 5**). The ML algorithms are trained using embedded atomic positions and their corresponding bond angles, which serve as local atomic environment descriptors. These features are labeled with associated forces and energies in the training dataset. In the literature, a diverse array of descriptors is employed for MLFF development. These encompass Behler–Parrinello vectors [93], atomic cluster expansion descriptors [94], Smooth Overlap of Atomic Positions (SOAP) [95], and graph-based descriptors [73], among others.

3.2 Machine-learning force field development for multi-component alloys

Creating MLFFs for single-element materials is significantly simpler compared to multi-component materials. This is attributed to the expansive composition space in multi-component materials, which grows substantially with the introduction of new elements into the system. This complexity renders the development of MLFFs challenging and relatively unexplored within the realm of computational material science. The challenge linked to the composition space stems from creating the training dataset, typically generated with DFT, which is an expensive process. Moreover, training a MLFF with all possible configurations covering the entire composition space, even if feasible, becomes computationally demanding due to the high dimensionality of atomic environment descriptors for multi-component configurations and constraints posed by memory limits [96–99]. Nevertheless, there have been efforts to develop MLFFs for materials with compositions involving up to five elements [90, 100–108], which will be discussed in the following section.

To tackle the dimensionality scaling challenge in MLFF development for multi-component alloys, the problem with SOAP descriptors is addressed by reducing the descriptor's dimensionality from quadratic to linear with respect to the number of elements [99]. The proposed methods include, firstly, considering the recoverability of density expansion coefficients from the power spectrum to create a compressed version that can be restored under specific conditions. Secondly, introducing a generalized SOAP kernel that enables compression concerning both the number of chemical elements and the number of radial basis functions in the density expansion. The performance of the compressed descriptors is then evaluated across various datasets using numerical tests that assess the accuracy of fitted energy models and MLFFs.

In a subsequent study, Byggmastar *et al.* [89, 90] discovered that augmenting the training data significantly for a GAP force-field with typical high-dimensional SOAP descriptors will not achieve satisfactory accuracy. They discovered that employing uncomplicated low-dimensional two-body and three-body descriptors in a GAP proved more effective than a SOAP-GAP for crystalline alloys. Additionally, this approach provided a noteworthy advantage, as a pure 2 + 3-body potential could be tabulated and evaluated efficiently using cubic-spline interpolations (referred to as tabGAP), resulting in a two-order-of-magnitude speed-up. The tabGAP model yielded results with near-quantum accuracy for a five-element Mo–Nb–Ta–V–W alloy, demonstrating the effectiveness of low-dimensional descriptors in handling complex alloy systems.

Byggmaster et al. [90] developed a MLFF designed for the Mo–Nb–Ta–V–W quinary system, employing it to investigate segregation and defects in the body-centered cubic (BCC) refractory high-entropy alloy MoNbTaVW in the GAP framework. They observed clear ordering in the bulk alloy, primarily involving Mo-Ta and V-W binaries at lower temperatures. Simulations in radiated crystals revealed pronounced segregation of V, the smallest atom in the alloy, to densely packed interstitial-rich regions like radiation-induced dislocation loops. The utility of MLFFs in modeling intricate multi-component alloys is evident, offering significant benefits to the material science community.

To assess the capability of MLFFs in accurately predicting alloy phase diagrams, Rosenbrock *et al.* [103] formulated an MLFF for Ag-Pd, aiming to characterize the energy of alloy configurations across a broad spectrum of compositions. The MLFF developed exhibits high accuracy across diverse compositions, demonstrating comparable results to the cluster expansion method [109], which serves as a reliable benchmark for this system. In a subsequent study, Xie *et al.* [79] created an MLFF for SiC polymorphs using the FLARE active learning framework, specifically designed to capture phase transformations in this material. Their work demonstrated strong agreement with both ab initio calculations and experimental measurements, surpassing existing empirical models in terms of vibrational and thermal properties. The examples mentioned above highlight the applicability of MLFFs in capturing diverse crystalline phases of a material, provided that the MLFF is developed accurately.

Marchand *et al.* [104] developed a series of NNIPs for the Al-Cu system, attaining near-first-principle accuracy across various metallurgically significant aspects. These

NNIPs accurately predict intermetallic compounds, elastic constants, solid-solution energetics, precipitate-matrix interfaces, stacking fault energies, and other properties. The machine learning approach, in comparison to state-of-the-art potentials, offers substantial quantitative benefits and exhibits promising transferability to defects and properties beyond the training structures, highlighting the crucial requirement for meticulous validation in specific metallurgical applications. In a subsequent study by Jain *et al.* [106], a set of NNIPs were created for Al-Mg-Si alloys, demonstrating significant applicability in the investigation of precipitate/matrix interfaces, antisite defect energies, and vacancy trapping in Mg-Si segregations. Furthermore, in another study by Marchand *et al.* [107], NNIPs were formulated for Al-Cu-Mg-Zn alloys. This research demonstrated that NNIPs could be effectively employed for more intricate systems, particularly in predicting plasticity-related properties at the quantum level.

Upon reviewing the literature on MLFF development for multi-component alloys, it becomes evident that despite significant progress in this field, there remains substantial work to be undertaken to enhance the effectiveness of MLFFs in alloy modeling across diverse aspects. While some strides have been made to extend the time and length scales of MLFFs, further improvements are needed, particularly for larger molecular dynamics simulations involving intricate nano-mechanical applications such as dislocation dynamics in nanoindentation and other nano-mechanical tests.

4. AI-assisted composition search and materials discovery

4.1 Machine learning and its role in accelerated search for compositions that exhibit specific functional properties in HEAs: An example for good practices

On the basis of the composition-property relationships, the expansive compositional flexibility afforded by multi-principal component alloys, such as High Entropy Alloys (HEAs), holds the potential to unlock a diverse array of material properties. An example of such potential is the so-called "cocktail" effect, which pertains to the synergistic interplay of various elements within high entropy compounds, giving rise to unexpected properties [110]. However, the advantages presented by such diversity hold a challenge in the disclosure of the optimal compositions for specific applications. In fact, navigating the vast landscape of HEAs compositions requires a departure from conventional trial-and-error approaches. The inherent inefficiency of such methods, both in terms of time and the uncertainty of practical utility, necessitates a more systematic exploration, especially considering the cost and complexity associated with experimental validations.

Density Functional Theory (DFT) [111] stands out as a highly precise means of simulating materials at the quantum level. However, its computational demands impose constraints on the scales of systems that can be effectively explored, rendering the comprehensive investigation of compositional spaces impractical. As a complementary tool, CALPHAD (Calculation of Phase Diagrams) [112] methods has proven to be precious in the context of HEAs [113–116] and the prediction of their phase structures, overcoming the limitations of the Hume-Rothery rules [117]. These methods utilize a thermodynamic approach, combining computational and experimental data to predict phase diagrams and thermodynamic properties. Nevertheless, empirical methods, including CALPHAD, may encounter challenges when extrapolating to compositions significantly divergent from the original databases [71].

By leveraging data-driven methodologies, Materials Informatics [22] accelerates computational experiments, offering a promising avenue for exploring compositional and functional spaces within materials. This integration of advanced computational tools holds immense potential for revolutionizing our understanding of material properties and streamlining the experimental validation process.

As an introduction to such methods, one could start from an example of a general procedure in the ML supported design of HEAs for specific target properties. C. Wen *et al.* [118], defined a virtuous procedure for the composition search on HEAs based on Al-Co-Cr-Cu-Fe-Ni to maximize their hardness, which will serve for us as a template for the explanation of ML property-oriented materials design, and the inherent good practices. The scheme of such procedure is presented in **Figure 6**. The first point of any ML application is the building of a high-quality dataset; this includes to pay particular attention to the source of the data, to data-cleaning and balancing techniques. In the mentioned work, the authors (i) collect samples in the dataset only if fabricated with the very same technique; (ii) remove outliers and samples with significantly different predictions of the target property in different works; (iii) validate with experiments the hardness measures of a small random subset of samples among the selected ones, to ensure small differences.

The next step, is the addition of supporting features for each sample. In a general Materials Informatics framework, such features can be anything from chemical compositions to structural or electronic properties, and it is crucial to understand the so-called *feature importance*, namely how relevant a set of features is for the sake of the ML model performance, and to remove redundant features. As it can be seen from the procedural scheme, C. Wen *et al.* analyzed the importance of features by separately

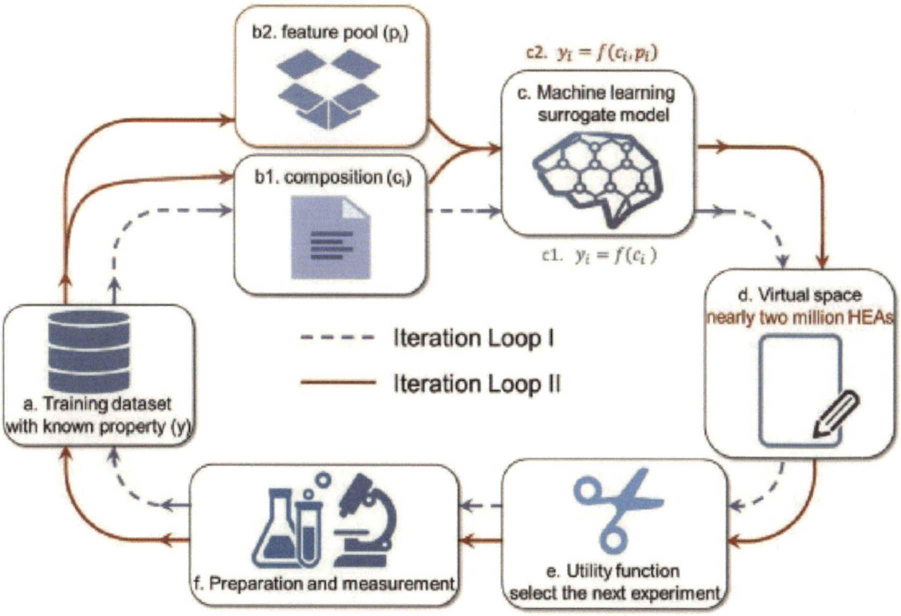

Figure 6.
Figure taken from [118]. Visualization of the property-oriented design of HEAs with the support of ML models. Good practices in the development of a materials discovery pipeline include: (a) a high-quality dataset; (b 1-2) materials features supporting the supervised learning with the ML model subsequently selected in (c); (d) the creation of the virtual space of potential compositions to be explored; (e) the definition of utility function of support in the selection of potential candidates for experiments; (f) experiments.

testing a ML model with and without materials knowledge, namely features beyond compositions that included phase formation (examples are atomic radii mismatch, mixing and configurational enthalpy, valence electron concentration, local electronegativity mismatch, cohesive energy) and strengthening theory and mechanical features (examples are modulus mismatch, Peierls-Nabarro factor, shear modulus difference).

Model selection is the crucial step in which the capabilities of different models to perform the regression task towards unseen materials properties is examined to promote the one with minimum prediction errors. After identifying the best ML model out of a set of candidates (i.e. linear, polynomial or support vector regression, or neural networks), it can be applied on the search space containing the large variety of samples (in C. Wen study case, unsynthesized alloys) with unknown target labels (hardness). During such explorations, C. Wen *et al.* have applied the precaution of maximizing, through the Efficient Global Optimization (EGO) [119] a utility function for the selection of the samples to be validated through experiments, the *expected improvement* function [120], defined as $EI = E[\max(y - \mu', 0)] = \sigma[\phi(z) + z\Phi(z)]$, where μ' is the maximum hardness value in the training dataset, and $\phi(z)$ and $\Phi(z)$ are the standard normal density and distribution functions of $z = (\mu - \mu')/\sigma$, with μ and σ the hardness and uncertainty predictions, respectively. The hardness experimental measures on the samples of confidence are then added to the training data, recalling what is known to be as *Active Learning*.

With such procedure, C. Wen *et al.*, found 35 out of 42 samples of confidence with higher hardness than the maximum contained in the training dataset. Clearly, each step of the process is and contains general enough practices that can be applied to any target property during HEAs composition searches and design.

In the next section, we present and discuss examples of how Machine learning can help exploring HEAs compositions for specific target properties, in particular focusing on energy and extreme environment applications.

4.2 AI-assisted accelerated search for compositions that exhibit specific functional properties in HEAs: Energy and environmental applications

The urgent transition towards green energy solutions clearly requires sophisticated solutions in terms of the target materials properties to be adapted for specific applications (i.e. automotive, solar, catalysis), but, in the first place, in terms of the searching strategies of materials displaying them. The actual exploration process which determines potential materials candidates a is time-consuming one, ramping the scales from the quantum-level investigation to the large scale validation of their characteristics. As introduced before, Machine Learning approaches are the most advanced tool that a scientist can deploy to speed up composition searches, and in this section the use of HEAs in energy applications is presented.

In the context of hydrogen storage, for automotive applications the US Department of Energy (D.O.E) identified the very strict criteria for characterizing a suitable storage systems in terms of costs, safety, reversibility, working thermodinamic conditions and more [121]. Metal hydrides [122] have been extensively studied as a potential solution, as they are able to absorb and release a significant amount of hydrogen through a chemical reaction, offering a safer option compared to high-pressure gas or liquid hydrogen storage [123–125], because they can operate at lower pressures [126] and are less prone to leakage. However, the development of efficient metal hydrides for hydrogen storage has been a persistent challenge For example, for hydrides primarily composed of transition metals, the problem is their constrained capacity, characterized by a low H/M ratio, while the acceptable storage capabilities of metal hydrides like the AB5-type, necessitate the incorporation of rare-earth metals like lanthanum, adding complexity and cost to the system [127]. While MgH2 and certain complex hydrides come close to meeting storage targets, their practical utility is hindered by challenges such as slow kinetics, limited reversibility, and high operating temperatures. These factors collectively render them unsuitable for applications with highly dynamic loads, such as those encountered in passenger cars [128].

In this context, HEAs represent a promising solution due to the high tunability of the compositions, to the advantageous capacities of BCC structures and the lattice distorsions due to the atomic radii mismatch between the constituent species [127]. As an example, from the point of view of BCC HEAs, $Ti_{0.325}V_{0.275}Zr_{0.125}Nb_{0.275}$ and TiVCrNb absorb hydrogen at room temperature, with moderate working pressure conditions, and are suitable for stationary hydrogen storage applications, while the intermetallic $Mg_{0.10}Ti_{0.30}V_{0.25}Zr_{0.10}Nb_{0.2}$ HEA composition reported promising hydrogen storage performance as it absorbed 1.7H/M (2.7 wt% H_2) at room temperature with fast kinetics [129].

A ML model for predicting thermodynamic properties of HEAs is an essential step to allow the fast screening of their potential for hydrogen storage applications. The reason lies in the thermodynamics basis of the HEA-H interaction, for a complete description of which we advice the reading of [129]. For our purposes, let us limit ourselves to mention the importance of the van't Hoff relation

$$ln\left(P_{eq}\right) = -\frac{\Delta H}{RT} + \frac{\Delta S}{R}, \tag{1}$$

which relates the equilibrium H_2 pressure at a given temperature T with the enthalpy and entropy of dissociation of the hydride phase. In a remarkable contribution, Witman *et al.* [131] trained a ML model (Gradient Boosting Tree) on the HydPARK metal hydrides database to predict such target feature, confirming the structure-property relationships intertwining ΔH and interstitial volumes, responsible for the dependence of the $P_{eq}(H_2)$ on the volume of the cell. The applied approach did not rely on the exact crystal structures of the systems in the dataset, making it a *compositional ML model* [132, 133], particularly suitable for the search of potential HEA hydrides yet to be synthesized. For this reason, in a later work, Witman *et al.* [132] based on the same approach the predictions of the hydride stability for a virtual space of 672 refractory HEAs (rHEAs) from the set of Al-Ti-V-Cr-Zr-Nb-Mo-Pd-Hf-Ta, after improving the model by adding reference rHEA samples in the dataset (TiVzrNbHf and TiVZrNb) with their known thermodynamic properties, as well as the binary hydride formation enthalpy per atom (given an element i), averaged and weighted by the composition fraction of the elementf_i: $\overline{\Delta H}_{bh} = \sum_i f_i \Delta H_{bh,i}$. Additionally, a predictive model for the stability of single phase solid solution for a given HEA composition was developed by training a classifier on the experimental database from Miracle *et al.* [71], and validating its predictions on a small dataset of synthesized compounds. Combining the predictions of the augmented model for the thermodynamic stability of rHEAs and the one for the formation of single-phase SS, the authors were able to identify the samples to be reproduced, characterized by improved destabilization (higher plateau pressures). In particular, AlTiVCr was identified to have a 70x increase in the plateau pressure, and conveniently included lighter-weight, lower-cost aluminum.

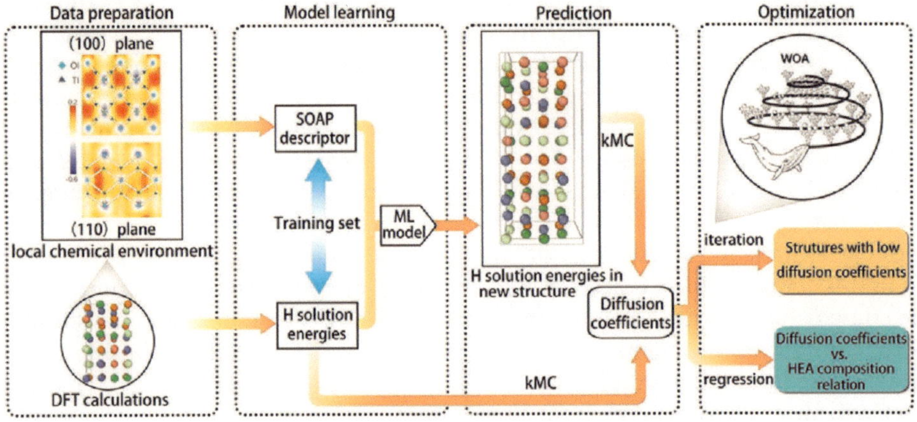

Figure 7.
Figure taken from [130]. Hydrogen solution energies are acquired by DFT simulations, and correlated to the SOAP descriptors of the local atomic environment. Predictions of hydrogen solution energies in unseen HEAs compositions, explored by the WOA algorithm, are then fed into kMC simulations to predict diffusion coefficients.

Another aspect inherent to the presence of hydrogen in metals, is the hydrogen-induce deterioration of their mechanical properties, also referred to as *hydrogen embrittlement* (HE), for example including reduction of tensile and fatigue strength, crack initiation and propagation [134]. For these reasons, such phenomena are of remarkable interest also for the hydrogen storage community. One way of mitigating the effects of HE, is to design materials with ultra-low hydrogen diffusion coefficients, in such a way to limit hydrogen accumulation at defects. X. Zhou *et al.* [130], define a machine learning supported procedure to design HE-resistant high-entropy alloys (HEAs), specifically focusing on FeCoNiCrMn compositions. Combining density functional theory to obtain hydrogen solution energies in the crystal structure, their use in kinetic Monte Carlo (kMC) simulations to access diffusion coefficients, and SOAP (Smooth Overlap of Atomic Positions) [135] descriptors for the local atomic environments of hydrogen, they target to highlight the intertwined relations via ML and to consequently lower the diffusion coefficients by tuning the HEAs compositions. In particular, the whale optimization algorithm (WOA) [136] is used to optimally explore the virtual HEAs compositional space, and by training a neural network to predict the hydrogen solution energy in unseen HEA SOAP descriptors, they can then be deployed within kMC simulations to obtain the diffusion coefficients. **Figure 7** reports such scheme of the procedure from their work. The model indicates that adjusting the alloy chemical composition, particularly elevating Co and Mn levels, enhances the FeCoNiCrMn alloys resilience against hydrogen embrittlement.

5. Conclusions

In this chapter, we have outlined the current state of materials informatics and its applications relevant to high-entropy alloys' composition, microstructure, and functionality. Our special focus has been on utilizing the machine learning toolbox pertinent to high-throughput materials design, micro-structural/chemical ordering quantification/tailoring, and search strategies for novel materials, offering useful insights into inherent processing-microstructure-property relationships in HEAs. The present overview has further addressed fundamental challenges concerning the highly dimensional and scattered nature of the parameter space associated with the high-throughput design of HEAs. We have also provided guidance on relevant mitigation strategies, offering a roadmap for machine-learning practitioners exploring the complexities of materials informatics in the context of HEAs.

Acknowledgements

The authors acknowledge funding from the European Union Horizon 2020 research and innovation program under Grant Agreement No. 857470 and from the European Regional Development Fund under the program of the Foundation for Polish Science International Research Agenda PLUS, grant No. MAB PLUS/2018/8, and the initiative of the Ministry of Science and Higher Education Support for the activities of Centers of Excellence established in Poland under the Horizon 2020 program' under agreement No. MEiN/2023/DIR/3795.

Author details

Amir H. Naghdi[1,2†], Dario Massa[1,2†], Kamran Karimi[1†] and Stefanos Papanikolaou[1*†]

1 NOMATEN Centre of Excellence, National Center for Nuclear Research, Otwock, Poland

2 IDEAS NCBR, Warsaw, Poland

*Address all correspondence to: stefanos.papanikolaou@ncbj.gov.pl

† These authors contributed equally.

References

[1] Pickering EJ, Jones NG. High-entropy alloys: A critical assessment of their founding principles and future prospects. International Materials Reviews. 2016;**61**(3):183-202

[2] Li Z, Zhao S, Ritchie RO, Meyers MA. Mechanical properties of high-entropy alloys with emphasis on face-centered cubic alloys. Progress in Materials Science. 2019;**102**:296-345

[3] George EP, Raabe D, Ritchie RO. High-entropy alloys. Nature Reviews Materials. 2019;**4**(8):515-534

[4] Shang Y, Brechtl J, Psitidda C, Liaw PK. Mechanical behavior of high-entropy alloys: A review. In: Brechtl J, Liaw PK, editors. High-entropy Materials: Theory, Experiments, and Applications. Cham: Springer; DOI: 10.1007/978-3-030-77641-1_10. arXiv preprint arXiv:2102.09055. 2021

[5] Jien-Wei YEH. Recent progress in high entropy alloys. Annales de Chimie Science des Materiaux. 2006;**31**(6): 633-648

[6] Zhang Y, Zhou YJ, Lin JP, Chen GL, Liaw PK. Solid-solution phase formation rules for multi-component alloys. Advanced Engineering Materials. 2008; **10**(6):534-538

[7] Li W, Wang G, Shiwei W, Liaw PK. Creep, fatigue, and fracture behavior of high-entropy alloys. Journal of Materials Research. 2018;**33**(19): 3011-3034

[8] Chen J, Zhou X, Wang W, Liu B, Lv Y, Yang W, et al. A review on fundamental of high entropy alloys with promising high–temperature properties. Journal of Alloys and Compounds. 2018; **760**:15-30

[9] Frydrych K, Karimi K, Pecelerowicz M, Alvarez R, Dominguez-Gutiérrez FJ, Rovaris F, et al. Materials informatics for mechanical deformation: A review of applications and challenges. Materials. 2021;**14**(19):5764

[10] Bierwolf R, Hohenstein M, Phillipp F, Brandt O, Crook GE, Ploog K. Direct measurement of local lattice distortions in strained layer structures by hrem. Ultramicroscopy. 1993;**49**(1–4): 273-285

[11] Hÿtch MJ, Snoeck E, Kilaas R. Quantitative measurement of displacement and strain fields from hrem micrographs. Ultramicroscopy. 1998;**74**(3):131-146

[12] Papanikolaou S, Alava MJ. Direct detection of plasticity onset through total-strain profile evolution. Physical Review Materials. 2021;**5**(8):083602

[13] Mäkinen T, Zaborowska A, Frelek-Kozak M, Jóźwik I, Kurpaska Ł, Papanikolaou S, et al. Detection of the onset of yielding and creep failure from digital image correlation. Physical Review Materials. 2022;**6**(10): 103601

[14] Papanikolaou S. Microstructural inelastic fingerprints and data-rich predictions of plasticity and damage in solids. Computational Mechanics. 2020; **66**:141-154

[15] Yang Z, Papanikolaou S, Reid ACE, Liao W-k, Choudhary AN, Campbell C, et al. Learning to predict crystal plasticity at the nanoscale: Deep residual networks and size effects in uniaxial compression discrete dislocation simulations. Scientific Reports. 2020; **10**(1):8262

[16] Chen X, Wang Q, Cheng Z, Zhu M, Zhou H, Jiang P, et al. Direct observation of chemical short-range order in a medium-entropy alloy. Nature. 2021; **592**(7856):712-716

[17] Fantin A, Lepore GO, Manzoni AM, Kasatikov S, Scherb T, Huthwelker T, et al. Short-range chemical order and local lattice distortion in a compositionally complex alloy. Acta Materialia. 2020;**193**:329-337

[18] Zhou L, Wang Q, Wang J, Chen X, Jiang P, Zhou H, et al. Atomic-scale evidence of chemical short-range order in crconi medium-entropy alloy. Acta Materialia. 2021;**224**:117490

[19] He QF, Tang PH, Chen HA, Lan S, Wang JG, Luan JH, et al. Understanding chemical short-range ordering/demixing coupled with lattice distortion in solid solution high entropy alloys. Acta Materialia. 2021;**216**:117140

[20] Miao J, Slone C, Dasari S, Ghazisaeidi M, Banerjee R, George EP, et al. Ordering effects on deformation substructures and strain hardening behavior of a crconi based medium entropy alloy. Acta Materialia. 2021;**210**: 116829

[21] Ziletti A, Kumar D, Scheffler M, Ghiringhelli LM. Insightful classification of crystal structures using deep learning. Nature Communications. 2018;**9**(1):2775

[22] Rickman JM, Chan HM, Harmer MP, Smeltzer JA, Marvel CJ, Roy A, et al. Materials informatics for the screening of multi-principal elements and high-entropy alloys. Nature Communications. 2019;**10**(1):2618

[23] Allen M, P, Tildesley DJ. Computer Simulation of Liquids. Oxford, UK: Oxford University Press; 2017

[24] Brandon D, Kaplan WD. Microstructural Characterization of Materials. England: John Wiley & Sons; 2013

[25] Ragone M, Shahabazian-Yassar R, Mashayek F, Yurkiv V. Deep learning modeling in microscopy imaging: A review of materials science applications. Progress in Materials Science. 2023;**138**: 101165

[26] Rao Z, Li Y, Zhang H, Colnaghi T, Marek A, Rampp M, et al. Direct recognition of crystal structures via three-dimensional convolutional neural networks with high accuracy and tolerance to random displacements and missing atoms. Scripta Materialia. 2023; **234**:115542

[27] Kaufmann K, Lane H, Liu X, Vecchio KS. Efficient few-shot machine learning for classification of ebsd patterns. Scientific Reports. 2021;**11**(1): 8172

[28] Liu S, Amin-Ahmadi B, Liu R, Zheng Q, Zhang X. Automated phase segmentation and quantification of high-resolution tem image for alloy design. Materials Characterization. 2023;**199**: 112779

[29] Frey NC, Akinwande D, Jariwala D, Shenoy VB. Machine learning-enabled design of point defects in 2d materials for quantum and neuromorphic information processing. ACS Nano. 2020;**14**(10):13406-13417

[30] Massa D, Cieśliński D, Naghdi A, Papanikolaou S. Substitutional alloying using crystal graph neural networks. AIP Advances. 2024;**14**(1):015023

[31] Massa D, Kaxiras E, Papanikolaou S. Alloy informatics through ab initio charge density profiles: Case study of hydrogen effects in face-centered cubic

crystals. Acta Materialia. 2024;**268**: 119773

[32] Kaufmann K, Zhu C, Rosengarten AS, Maryanovsky D, Harrington TJ, Marin E, et al. Crystal symmetry determination in electron diffraction using machine learning. Science. 2020;**367**(6477): 564-568

[33] Spannaus A, Law KJH, Luszczek P, Nasrin F, Micucci CP, Liaw PK, et al. Materials fingerprinting classification. Computer Physics Communications. 2021;**266**:108019

[34] Spannaus A. Advanced statistical methods for Atomic-level quantification of multi-component alloys. PhD diss, University of Tennessee. 2020. Available from: https://trace.tennessee.edu/utk_graddiss/5889

[35] Wolverton C, Ozolins V, Zunger A. Short-range-order types in binary alloys: A reflection of coherent phase stability. Journal of Physics: Condensed Matter. 2000;**12**(12):2749

[36] Yuan W, Zhang F, Yuan X, Huang H, Wen X, Wang Y, et al. Short-range ordering and its effects on mechanical properties of high-entropy alloys. Journal of Materials Science & Technology. 2021;**62**:214-220

[37] Zhou L, Wang Q, Wang J, Chen X, Jiang P, Zhou H, et al. Atomic-scale evidence of chemical short-range order in crconi medium-entropy alloy. Acta Materialia. 2022;**224**:117490

[38] Wang J, Jiang P, Yuan F, Xiaolei W. Chemical medium-range order in a medium-entropy alloy. Nature Communications. 2022;**13**(1):1021

[39] Li Q-J, Sheng H, Ma E. Strengthening in multi-principal element alloys with local-chemical-order roughened dislocation pathways. Nature Communications. 2019;**10**(1): 1-11

[40] Naghdi AH, Karimi K, Poisvert AE, Esfandiarpour A, Alvarez R, Sobkowicz P, et al. Dislocation plasticity in equiatomic nicocr alloys: Effect of short-range order. Physical Review B. 2023;**107**(9):094109

[41] Yang X, Xi Y, He C, Chen H, Zhang X, ShanTung T. Chemical short-range order strengthening mechanism in cocrni medium-entropy alloy under nanoindentation. Scripta Materialia. 2022;**209**:114364

[42] Zhang R, Zhao S, Ding J, Chong Y, Jia T, Ophus C, et al. Short-range order and its impact on the crconi medium-entropy alloy. Nature. 2020;**581**(7808): 283-287

[43] Ding Q, Zhang Y, Chen X, Xiaoqian F, Chen D, Chen S, et al. Tuning element distribution, structure and properties by composition in high-entropy alloys. Nature. 2019;**574**(7777): 223-227

[44] Seol JB, Ko W-S, Sohn SS, Na MY, Chang HJ, Heo Y-U, et al. Mechanically derived short-range order and its impact on the multi-principal-element alloys. Nature Communications. 2022;**13**(1): 6766

[45] Naghdi A, Dominguez-Gutierrez FJ, Huo WY, Karimi K, Papanikolaou S. Dynamic nanoindentation and short-range order in equiatomic NiCoCr medium-entropy alloy lead to novel density wave ordering. Physical Review Letters. 2024;**132**(11):116101

[46] Esfandiarpour A, Alvarez-Donado R, Papanikolaou S, Alava M. Atomistic

simulations of dislocation plasticity in concentrated vconi medium entropy alloys: Effects of lattice distortion and short range order. Frontiers in Materials. 2022;**9**:1046291

[47] Jian W-R, Xie Z, Shuozhi X, Yanqing S, Yao X, Beyerlein IJ. Effects of lattice distortion and chemical short-range order on the mechanisms of deformation in medium entropy alloy CoCrNi. Acta Materialia. 2020;**199**: 352-369

[48] Basu I, De Hosson JTM. Strengthening mechanisms in high entropy alloys: Fundamental issues. Scripta Materialia. 2020;**187**:148-156

[49] Esfandiarpour A, Papanikolaou S, Alava M. Edge dislocations in multicomponent solid solution alloys: Beyond traditional elastic depinning. Physical Review Research. 2022;**4**(2): L022043

[50] Karimi K, Esfandiarpour A, Papanikolaou S. Serrated plastic flow in slowly-deforming complex concentrated alloys: Universal signatures of dislocation avalanches. DOI: 10.48550/ arXiv.2310.03828

[51] Schwartz AJ, Kumar M, Adams BL, Field DP. Electron Backscatter Diffraction in Materials Science. Vol. 2. NewYork, NY: Springer; 2009

[52] Randle V, Engler O. Introduction to Texture Analysis: Macrotexture, Microtexture and Orientation Mapping. London, UK: CRC Press; 2000

[53] Wilkinson AJ, Clarke EE, Britton TB, Littlewood P, Karamched PS. High-resolution electron backscatter diffraction: An emerging tool for studying local deformation. The Journal of Strain Analysis for Engineering Design. 2010;**45**(5):365-376

[54] Beyerlein IJ, Capolungo L, Marshall PE, McCabe RJ, Tomé CN. Statistical analyses of deformation twinning in magnesium. Philosophical Magazine. 2010;**90**(16):2161-2190

[55] Zhu C, Harrington T, Gray III GT, Vecchio KS. Dislocation-type evolution in quasi-statically compressed polycrystalline nickel. Acta Materialia. 2018;**155**:104-116

[56] Salmenjoki H, Papanikolaou S, Shi D, Tourret D, Cepeda-Jiménez CM, Pérez-Prado MT, et al. Machine learning dislocation density correlations and solute effects in mg-based alloys. Scientific Reports. 2023;**13**(1):11114

[57] Mulewska K, Dominguez-Gutierrez FJ, Kalita D, Jesper Byggmästar GY, Wei W, et al. Self–ion irradiation of high purity iron: Unveiling plasticity mechanisms through nanoindentation experiments and large-scale atomistic simulations. Journal of Nuclear Materials. 2023;**586**:154690

[58] Domínguez-Gutíerrez FJ, Ustrzycka A, Xu QQ, Alvarez-Donado R, Papanikolaou S, Alava M. Dislocation nucleation mechanisms during nanoindentation of concentrated FeNiCr alloys: Unveiling the effects of Cr through molecular simulations. Modelling Simulation Material Science Engineering. 2022;**30**: 085010

[59] Karimi K, Salmenjoki H, Mulewska K, Kurpaska L, Kosińska A, Alava MJ, et al. Prediction of steel nanohardness by using graph neural networks on surface polycrystallinity maps. Scripta Materialia. 2023;**234**: 115559

[60] Damewood J, Karaguesian J, Lunger JR, Tan AR, Xie M, Peng J, et al.

Representations of materials for machine learning. Annual Review of Materials Research. 2023;**53**:399-426

[61] Himanen L, Jäger MOJ, Morooka EV, Canova FF, Ranawat YS, Gao DZ, et al. Dscribe: Library of descriptors for machine learning in materials science. Computer Physics Communications. 2020;**247**:106949

[62] Behler J. Atom-centered symmetry functions for constructing high-dimensional neural network potentials. The Journal of Chemical Physics. 2011; **134**(7):074106

[63] Goryaeva AM, Lapointe C, Dai C, Dérès J, Maillet J-B, Marinica M-C. Reinforcing materials modelling by encoding the structures of defects in crystalline solids into distortion scores. Nature Communications. 2020;**11**(1): 4691

[64] Grigorev P, Goryaeva AM, Marinica M-C, Kermode JR, Swinburne TD. Calculation of dislocation binding to helium-vacancy defects in tungsten using hybrid ab initio-machine learning methods. Acta Materialia. 2023;**247**:118734

[65] Lapointe C, Swinburne TD, Proville L, Becquart CS, Mousseau N, Marinica M-C. Machine learning surrogate models for strain-dependent vibrational properties and migration rates of point defects. Physical Review Materials. 2022;**6**(11):113803

[66] Malerba L, Caturla MJ, Emile Gaganidze C, Kaden MJK, Olsson P, Robertson C, et al. Multiscale modelling for fusion and fission materials: The m4f project. Nuclear Materials and Energy. 2021;**29**:101051

[67] Swinburne TD. Uncertainty and anharmonicity in thermally activated dynamics. Computational Materials Science. 2021;**193**:110256

[68] Goryaeva AM, Dérès J, Lapointe C, Grigorev P, Swinburne TD, Kermode JR, et al. Efficient and transferable machine learning potentials for the simulation of crystal defects in bcc fe and w. Physical Review Materials. 2021;**5**(10):103803

[69] Jakse N, Sandberg J, Granz LF, Saliou A, Jarry P, Devijver E, et al. Machine learning interatomic potentials for aluminium: Application to solidification phenomena. Journal of Physics: Condensed Matter. 2022;**51**(3): 035402

[70] Freitas R, Cao Y. Machine-learning potentials for crystal defects. MRS Communications. 2022;**12**(5): 510-520

[71] Miracle DB, Senkov ON. A critical review of high entropy alloys and related concepts. Acta Materialia. 2017;**122**: 448-511

[72] Karimi K, Papanikolaou S. Multiscale modeling of kinetic sluggishness in equiatomic nicocr and nicocrfemn single-phase solid solutions. DOI: 10.48550/arXiv.2304.04255

[73] Musaelian A, Batzner S, Johansson A, Sun L, Owen CJ, Kornbluth M, et al. Learning local equivariant representations for large-scale atomistic dynamics. Nature Communications. 2023;**14**(1):579

[74] Owen LR, Pickering EJ, Playford HY, Stone HJ, Tucker MG, Jones NG. An assessment of the lattice strain in the crmnfeconi high-entropy alloy. Acta Materialia. 2017;**122**:11-18

[75] Owen CJ, Torrisi SB, Xie Y, Batzner S, Bystrom K, Coulter J, et al. Complexity of Many-Body Interactions

in Transition Metals Via Machine-Learned Force Fields from the TM23 Data Set; 2023. arXiv preprint arXiv: 2302.12993

[76] Owen CJ, Xie Y, Johansson A, Sun L, Kozinsky B. Stability, mechanisms and kinetics of emergence of Au surface reconstructions using Bayesian force fields. 2023. arXiv preprint arXiv: 2308.07311

[77] Owen CJ, Marcella N, Xie Y, Vandermause J, Frenkel AI, Nuzzo RG, Kozinsky B. Unraveling the Catalytic Effect of Hydrogen Adsorption on pt Nanoparticle Shape-Change. 2023. arXiv preprint arXiv:2306.00901

[78] Phuthi MK, Yao AM, Batzner S, Musaelian A, Kozinsky B, Cubuk ED, Viswanathan V. Accurate Surface and Finite-Temperature Bulk Properties of Lithium Metal at Large Scales using Machine Learning Interaction Potentials. ACS Omega. 2024;**9**(9):10904-10912

[79] Xie Y, Vandermause J, Ramakers S, Protik NH, Johansson A, Kozinsky B. Uncertainty-aware molecular dynamics from bayesian active learning for phase transformations and thermal transport in sic. npj Computational Materials. 2023;**9**(1):36

[80] Batzner S, Musaelian A, Sun L, Geiger M, Mailoa JP, Kornbluth M, et al. E(3)-equivariant graph neural networks for data-efficient and accurate interatomic potentials. Nature Communications. 2022;**13**(1):2453

[81] Vandermause J, Torrisi SB, Simon Batzner Y, Xie LS, Kolpak AM, Kozinsky B. On-the-fly active learning of interpretable bayesian force fields for atomistic rare events. npj Computational Materials. 2020;**6**(1):1-11

[82] Pellegrini F, Lot R, Shaidu Y, Küçükbenli E. Panna 2.0: Efficient Neural Network Interatomic Potentials and New Architectures. 2023

[83] Shaidu Y, Küçükbenli E, Lot R, Pellegrini F, Kaxiras E, de Gironcoli S. A systematic approach to generating accurate neural network potentials: The case of carbon. npj Computational Materials. 2021;**7**(1):52

[84] Lot R, Pellegrini F, Shaidu Y, Ksüçükbenli E. Panna: Properties from artificial neural network architectures. Computer Physics Communications. 2020;**256**:107402

[85] Li Q-J, Küçükbenli E, Lam S, Khaykovich B, Kaxiras E, Li J. Development of robust neural-network interatomic potential for molten salt. Cell Reports Physical Science. 2021;**2**(3): 100359

[86] Jonathan Vandermause Y, Xie JS, Lim CJ, Owen, and Boris Kozinsky. Active learning of reactive bayesian force fields applied to heterogeneous catalysis dynamics of h/pt. Nature Communications. 2022; **13**(1):5183

[87] Naghdi AD, Pellegrini F, Küçükbenli E, Massa D, Gutierrez FJD, Kaxiras E, Papanikolaou S. Neural Network Interatomic Potentials for Open Surface Nano-Mechanics Applications, 2023

[88] Bartók AP, Payne MC, Kondor R, Csányi G. Gaussian approximation potentials: The accuracy of quantum mechanics, without the electrons. Physical Review Letters. 2010;**104**:136403

[89] Byggmästar J, Nordlund K, Djurabekova F. Simple machine-learned interatomic potentials for complex alloys. Physical Review Materials. 2022; **6**:083801

[90] Byggmästar J, Nordlund K, Djurabekova F. Modeling refractory high-entropy alloys with efficient machine-learned interatomic potentials: Defects and segregation. Physical Review B. 2021;**104**:104101

[91] Wen M, Tadmor EB. Uncertainty quantification in molecular simulations with dropout neural network potentials. npj Computational Materials. 2020;**6**(1): 124

[92] Deringer VL, Bartók AP, Bernstein N, Wilkins DM, Ceriotti M, Csányi G. Gaussian process regression for materials and molecules. Chemical Reviews. 2021;**121**(16): 10073-10141

[93] Behler J, Parrinello M. Generalized neural-network representation of high-dimensional potential-energy surfaces. Physical Review Letters. 2007;**98**: 146401

[94] Drautz R. Atomic cluster expansion for accurate and transferable interatomic potentials. Physical Review B. 2019;**99**: 014104

[95] Bartók AP, Kondor R, Csányi G. On representing chemical environments. Physical Review B. 2013;**87**:184115

[96] Artrith N, Urban A, Ceder G. Efficient and accurate machine-learning interpolation of atomic energies in compositions with many species. Physical Review B. 2017;**96**:014112

[97] Imbalzano G, Anelli A, Giofré D, Klees S, Behler J, Ceriotti M. Automatic selection of atomic fingerprints and reference configurations for machine-learning potentials. The Journal of Chemical Physics. 2018;**148**(24):241730

[98] Willatt MJ, Musil F, Ceriotti M. Feature optimization for atomistic machine learning yields a data-driven construction of the periodic table of the elements. Physical Chemistry Chemical Physics. 2018;**20**:29661-29668

[99] Darby JP, Kermode JR, Csányi G. Compressing local atomic neighbourhood descriptors. npj Computational Materials. 2022; **8**(1):166

[100] Nikoulis G, Byggmästar J, Kioseoglou J, Nordlund K, Djurabekova F. Machine-learning interatomic potential for w–mo alloys. Journal of Physics: Condensed Matter. 2021;**33**(31):315403

[101] Koskenniemi M, Byggmästar J, Nordlund K, Djurabekova F. Efficient atomistic simulations of radiation damage in w and w–mo using machine-learning potentials. Journal of Nuclear Materials. 2023;**577**:154325

[102] Hajinazar S, Shao J, Kolmogorov AN. Stratified construction of neural network based interatomic models for multicomponent materials. Physical Review B. 2017;**95**:014114

[103] Rosenbrock CW, Gubaev K, Shapeev AV, Pártay LB, Bernstein N, Csányi G, et al. Machine-learned interatomic potentials for alloys and alloy phase diagrams. npj Computational Materials. 2021;**7**(1):24

[104] Marchand D, Jain A, Glensk A, Curtin WA. Machine learning for metallurgy i. a neural-network potential for al-cu. Physical Review Materials. 2020;**4**:103601

[105] Stricker M, Yin B, Mak E, Curtin WA. Machine learning for metallurgy ii. A neural-network potential for magnesium. Physical Review Materials. 2020;**4**:103602

[106] Jain ACP, Marchand D, Glensk A, Ceriotti M, Curtin WA. Machine learning for metallurgy iii: A neural network potential for al-mg-si. Physical Review Materials. 2021;**5**:053805

[107] Marchand D, Curtin WA. Machine learning for metallurgy iv: A neural network potential for al-cu-mg and al-cu-mg-zn. Physical Review Materials. 2022;**6**:053803

[108] Liyanage M, Reith D, Eyert V, Curtin WA. Machine learning for metallurgy v: A neural-network potential for zirconium. Physical Review Materials. 2022;**6**:063804

[109] Qu W, He B, Song T, Gao J, Shi S. Cluster expansion method and its application in computational materials science. Computational Materials Science. 2016;**125**:243-254

[110] Cao BX, Wang C, Yang T, Liu CT. Cocktail effects in understanding the stability and properties of face-centered-cubic high-entropy alloys at ambient and cryogenic temperatures. Scripta Materialia. 2020; **187**:250-255

[111] Kohn W, Sham LJ. Self-consistent equations including exchange and correlation effects. Physical review. 1996;**140**(4A):A1133

[112] Saunders N, Peter A, Miodownik. CALPHAD (Calculation of Phase Diagrams): A Comprehensive Guide. New York, NY, USA: Elsevier; 1998

[113] Feng R, Zhang C, Gao MC, Pei Z, Zhang F, Chen Y, et al. High-throughput design of high-performance lightweight high-entropy alloys. Nature Communications. 2021;**12**(1): 4329

[114] Li T, Wang S, Fan W, Yiping L, Wang T, Li T, et al. Calphad-aided design for superior thermal stability and mechanical behavior in a tizrhfnb refractory high-entropy alloy. Acta Materialia. 2023;**246**:118728

[115] Yang S, Jun L, Xing F, Zhang L, Zhong Y. Revisit the vec rule in high entropy alloys (heas) with high-throughput calphad approach and its applications for material design-a case study with al–co–cr–fe–ni system. Acta Materialia. 2020;**192**:11-19

[116] Reynolds CR, Herl Z, Ley NA, Choudhuri D, Lloyd JT, Young ML. Comparing calphad predictions with high energy synchrotron radiation x-ray diffraction measurements during in situ annealing of al0. 3cocrfeni high entropy alloy. Materialia. 2020;**12**: 100784

[117] Wan X, Li Z, Wei Y, Wang A, Ke X, Guo H, et al. Machine learning paves the way for high entropy compounds exploration: Challenges, progress, and outlook. Advanced Materials. 2023;**2023**: 2305192

[118] Wen C, Zhang Y, Wang C, Xue D, Bai Y, Antonov S, et al. Machine learning assisted design of high entropy alloys with desired property. Acta Materialia. 2019;**170**:109-117

[119] Ghiringhelli LM, Vybiral J, Levchenko SV, Draxl C, Scheffler M. Big data of materials science: Critical role of the descriptor. Physical Review Letters. 2015;**114**(10):105503

[120] Xue D, Balachandran PV, Hogden J, Theiler J, Xue D, Lookman T. Accelerated search for materials with targeted properties by adaptive design. Nature Communications. 2016;7(1):1-9

[121] O'Malley K, Ordaz G, Adams J, Randolph K, Ahn CC, Stetson NT.

Applied hydrogen storage research and development: A perspective from the US Department of Energy. Journal of Alloys and Compounds. 2015;**645**:S419-S422

[122] Von Colbe JB, Ares J-R, Barale J, Baricco M, Buckley C, Capurso G, et al. Application of hydrides in hydrogen storage and compression: Achievements, outlook and perspectives. International Journal of Hydrogen Energy. 2019; **44**(15):7780-7808

[123] Sun Y, Shen C, Lai Q, Liu W, Wang D-W, Aguey-Zinsou K-F. Tailoring magnesium based materials for hydrogen storage through synthesis: Current state of the art. Energy Storage Materials. 2018;**10**:168-198

[124] Felderhoff M, Weidenthaler C, von Helmolt R, Eberle U. Hydrogen storage: The remaining scientific and technological challenges. Physical Chemistry Chemical Physics. 2007;**9**: 2643-2653

[125] Hua TQ, Ahluwalia RK, Peng J-K, Kromer M, Lasher S, McKenney K, et al. Technical assessment of compressed hydrogen storage tank systems for automotive applications. International Journal of Hydrogen Energy. 2011;**36**(4): 3037-3049

[126] Aceves SM, Berry GD, Martinez-Frias J, Espinosa-Loza F. Vehicular storage of hydrogen in insulated pressure vessels. International Journal of Hydrogen Energy. 2006;**31**(15): 2274-2283

[127] Sahlberg M, Karlsson D, Zlotea C, Jansson U. Superior hydrogen storage in high entropy alloys. Scientific Reports. 2016;**6**(1):36770

[128] Klopčič N, Grimmer I, Winkler F, Sartory M, Trattner A. A review on metal hydride materials for hydrogen storage. Journal of Energy Storage. 2023; **72**:108456

[129] Marques F, Balcerzak M, Winkelmann F, Zepon G, Felderhoff M. Review and outlook on high-entropy alloys for hydrogen storage. Energy & Environmental Science. 2021;**14**(10): 5191-5227

[130] Zhou X-Y, Zhu J-H, Yuan W, Yang X-S, Lookman T, Hong-Hui W. Machine learning assisted design of feconicrmn high-entropy alloys with ultra-low hydrogen diffusion coefficients. Acta Materialia. 2022;**224**:117535

[131] Witman M, Ling S, Grant DM, Walker GS, Agarwal S, Stavila V, et al. Extracting an empirical intermetallic hydride design principle from limited data via interpretable machine learning. The Journal of Physical Chemistry Letters. 2019;**11**(1):40-47

[132] Witman M, Ek G, Ling S, Chames J, Agarwal S, Wong J, et al. Data-driven discovery and synthesis of high entropy alloy hydrides with targeted thermodynamic stability. Chemistry of Materials. 2021;**33**(11):4067-4076

[133] Ward L, Agrawal A, Choudhary A, Wolverton C. A general-purpose machine learning framework for predicting properties of inorganic materials. npj Computational Materials. 2016;**2**(1):1-7

[134] Li X, Ma X, Zhang J, Akiyama E, Wang Y, Song X. Review of hydrogen embrittlement in metals: Hydrogen diffusion, hydrogen characterization, hydrogen embrittlement mechanism and prevention. Acta Metallurgica Sinica (English Letters). 2020;**33**:759-773

[135] Csányi G, Willatt MJ, Ceriotti M. Machine-learning of atomic-scale

properties based on physical principles. In: Schütt K, Chmiela S, von Lilienfeld O, Tkatchenko A, Tsuda K, Müller KR, editors. Machine Learning Meets Quantum Physics. Cham: Springer; 2020;**968**:99-127. DOI: 10.1007/978-3-030-40245-7_6

[136] Mirjalili S, Lewis A. The whale optimization algorithm. Advances in Engineering Software. 2016;**95**:51-67

Chapter 3

Machine Learning Guides the Discovery of High-Performance HEA Catalysts

Jike Wang, Min Wei and Junyu Zhang

Abstract

High performance catalysts are crucial to generating clean fuels, reducing the impact of global warming, and providing solutions to environmental pollution. Improved processes for catalyst design and a better understanding of catalytic processes are key for improving the effectiveness and activities. HEAs typically have at least four principal elements, this atomic structure gives them unique properties that have applications and excellent performance in a variety of fields including catalysis. The complexity of HEAs makes challenge for computational researchers, providing promising opportunities for the application of machine learning. Recent advances in data science have great potential to accelerate catalyst research, particularly the rapid exploration of large materials chemistry spaces through machine learning. Here a comprehensive and critical review of machine learning techniques used in HEA catalysis research is provided. Sources of HEA catalyst data and current approaches to represent these materials by mathematical features are described, the most commonly used machine learning methods summarized, and the quality and utility of catalyst models evaluated. Illustrations of how machine learning models are applied to novel HEA catalysts discovery and used to reveal catalytic reaction mechanisms are provided.

Keywords: machine learning, high entropy alloy, high performance catalysts, electrocatalysis, photocatalysis

1. Introduction

High-entropy alloys (HEAs) have attracted tremendous attention due to the unique properties since the first reported (**Figure 1**) [1–3]. There are composition and entropy aspects of the HEA definition. Typically, HEAs have at least four principal elements with and the concentration of constituent metals would range from 5 at% to 35 at% in composition-based definition [4]. Meanwhile, according to Boltzmann's formula, the entropy of alloy is relevant to a given state, the relationship is following the Eq. (1):

$$s = k_B ln\omega \tag{1}$$

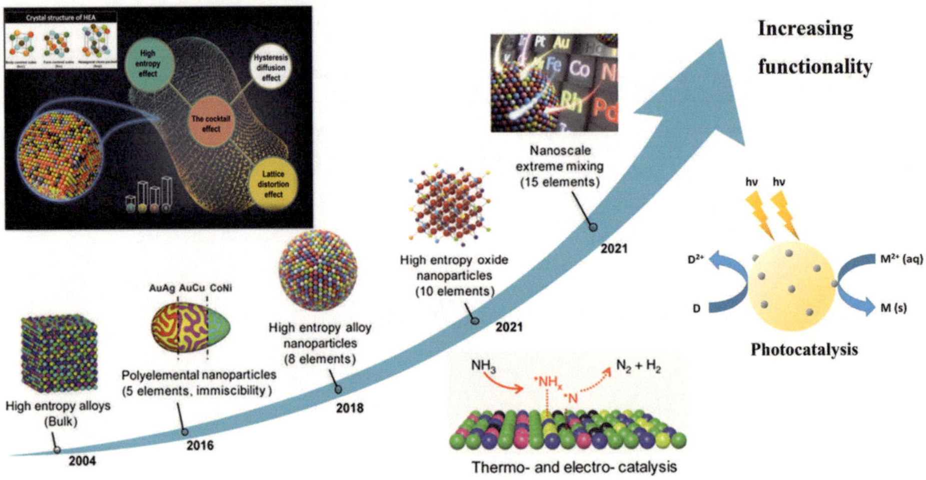

Figure 1.
The development of crystal structure and four major effects of HEAs [1]; development of HEAs [2], the increasing functionality in thermos/electro-catalysis and photocatalysis [3].

where k_B and ω represent the Boltzmann constant and the probability that a given state exists. The entropy of an alloy mainly consists of configurational entropy (S_{conf}), vibrational entropy, magnetic entropy and electronic randomness entropy [1]. Among these entropies, conformational entropy plays a major contribution in HEAs due to the fact that it is the primary driver of single-phase mixing. Considering the maximization of the conformational entropy of the alloy, the equimolar ratios of the constituent elements are selected. Thus, the mixing entropy of a system would be expressed as the following Eq. (2):

$$S_{conf} = Rln(n) \tag{2}$$

where R is the molar gas constant (R = 8.314 J·mol·K^{-1}) and n is the number of equimolar elements [5]. Therefore, the alloy with $S_{conf} \geq 1.5$ R containing the number of elemental components ≥5 is identified as a HEA [6, 7]. These two definitions, based on composition and entropy, encompass a wide range of high-entropy alloys and overlap in most cases.

As a result of the mixing entropy and multi-element composition of HEAs, the high-entropy effect, slow diffusion effect, cocktail effect, and lattice distortion effect are generated, which result in excellent physicochemical properties. And the relationship between these effects and properties are summarized as following (**Figure 1**): (i) the high-entropy effect refers to the enhanced configuration entropy in alloys originating from the increase of the number of components. The enhanced entropy of the mixing configuration facilitates the formation of a stable single-phase solid solution and prevents degradation process [8], thus contributing directly to the excellent stability of HEAs. (ii) The lattice distortion effect generated by the differences of atomic sizes, coordinate structure and electronic distributions among the constituent elements. Random atomic occupation leads to significant lattice distortion that modifies the electronic structure of the active metal atoms, such as the d-band center and the strain effect, thus resulting in higher potential energies and lowering the energy barriers for the catalytic reaction. (iii) The lattice distortion in HEAs would increase the diffusion activation energy and thus exhibit the slow diffusion effect, which is benefit to form the nanoscale HEAs on the surface in the high temperature and corrosion. (iv) The cocktail effect refers to the fact that synergistic interactions between the various constituent elements will have an unexpected property. The cocktail effect of HEAs is unpredictable rather than a stacking of the constituent metal elements. And the complex catalytic reactions facilitated by multiple active sites of HEAs in catalysis. In a word, the lattice distortion modifies the electronic structure of the active metals, and the high-entropy effect and slow diffusion effect enable the thermodynamic and kinetic stability of HEAs [9].

It's acknowledged that the configuration has great effect on the catalytic performance. A promising opportunity and the major challenge in the field of HEA catalysis is the abundance of possible compositions. For the screening of high-performance HEA catalysts, it is impracticable to rely on artificial experimental synthesis due to the structural complexity (configuration, composition and atomic ratio), which makes challenge for computational researchers and thus providing promising opportunities for the application of machine learning (ML). Recent advances in data science have great potential to accelerate catalyst research, particularly the rapid exploration of large materials chemistry spaces through ML.

2. ML in catalysis

2.1 Workflow for ML-based catalysts discovery and design

A general workflow for catalysts discovery and design shown in **Figure 2** involves four main steps: (i) datasets, (ii) descriptors, (iii) algorithm, and (iv) model evaluation and optimization. The first step is to obtain suitable target datasets from open and shared databases or data generated from experiments and simulations. The second step is to build a descriptor of the target data by collecting or building features related to the predicted target. The third step is to construct a model between descriptors and target attributes based on statistical methods such as simple linear regression, nonlinear regression, highly complex kernel ridge regression, and neural networks. In the last step, the generalization error of the model is evaluated by a computation-based test and optimized based on the evaluation results.

2.2 Data collection

Data provides the samples that are used to train and test the model in ML. The quality and quantity of data have a decisive impact on the performance of ML. If the data quality is poor, the machine learning algorithm will not be able to learn effective patterns and regularities from it, which will affect its performance. And the amount of data needs to be determined according to the complexity of the specific problem and the limitations of the available resources.

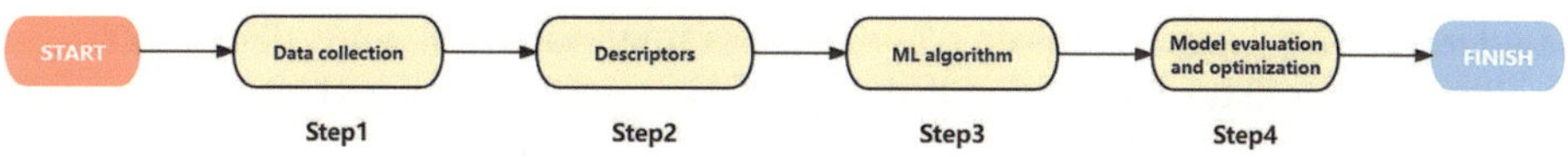

Figure 2.
The workflow for ML-based material discovery and design.

The data mainly originates from databases, experiments, and first-principles calculation. 370 HEAs containing complex concentrated mechanical properties were established in 2018 [10]. In general, the data obtained from the experiment can be directly used for training of the model due to its accuracy. Using experimental data exclusively as a dataset is impractical due to the high time-consumption. Thus, the emergence of DFT calculations compensates for the limitations of experimentation. The first-principles approach, which is based on quantum mechanics and whose computational process involves atomic species and positional coordinates, has become the choice for catalysts design and exploration. DFT calculations can be used to generate large datasets and perform model training. And this method is adopted by most researchers to study the catalytic performance of HEA under different reaction types such as CO_2RR [11], OER [12], ORR [13], NH_3OR [14].

2.3 Descriptors

It's hard to understand the properties of catalyst with only dataset, and descriptors are essential to improve the accuracy of models [15]. Descriptors are generally repeatable, valid, distinguishable and simple. Repeatability means that the descriptors of the same material are fixed to ensure the reproducibility of machine learning results. The descriptors should contain at least one intrinsic property of the catalyst, and the different catalyst possesses the distinguish descriptors. Thus, catalyst that can theoretically be distinguished by descriptors. Additionally, the simplicity of the descriptors effectively improves the accuracy of the model.

Generally, descriptors are classified into electronic structure, geometric structure, and adsorption energy [16, 17], as the geometric structure including atomic radius [18], atomic number [19, 20], molar volume [21], lattice constants [22], rotational angle [23], bonding information [24], coordination number [25] illustrates the adsorption of intermediates, which affects the catalytic activity. Meanwhile, the electronic structure such as valence electrons [26], electronegativity [27], electron affinity, etc. reflects the electron transfer direction and charge distribution. In addition, the adsorption energy of active site is the direct evidence to judge the catalytic activity [28]. In general, the microstructural features of the adsorption sites used to describe the geometrical characteristics of the materials and the adsorption energy used to describe the catalytic activity are the most common choice of descriptors.

2.4 ML algorithm

Machine learning algorithm is critical in prediction of high-activity catalysts, whose performance is largely relied on the size of the data and the accuracy of designed model. In addition, model selection is most important step in the very beginning of the prediction process, since every algorithm has its own advantage and limitation. A proper model should be selected after a comprehensive consideration of multiple situations, including data amount, data format, application scenario and so on. For example, traditional ML algorithms may not have a worse performance than deep learning algorithms do when process small samples, while deep learning algorithms will do a better job if the sample amount is sufficient enough due to their strong learning ability, wide application range and excellent performance for complexity. Therefore, in order to achieve a reasonable selection, a comprehensive summary of the most common ML algorithms is presented.

Linear regression algorithm is the most commonly employed algorithm in ML owing to the understandability, computational efficiency, and interpretability [29]. This algorithm is mainly for establishing a linear relationship between the independent and dependent variables, making predictions and analyzing results. Least squares algorithm is the most widely practiced optimization algorithm in linear regression [30]. By minimizing the sum of squared errors to find the best function that matches the data, and quickly obtain the unknown data, minimize the sum of squared errors between the obtained data and the actual data. It is an unbiased estimation method with the advantage of being computationally simplicity and stability, which is suitable for the case where there is not excessive noise in the data and there is no covariance between the data. The results of the least squares method may be unstable when covariance exists, even if the noise is weak. In addition, the least squares method has no parameter selection for variable selection or sparsity.

Kernel methods [31] including Support Vector Machine (SVM) [32], Gaussian Process Regression (GPR) [33], and Kernel Ridge Regression (KRR) [34], are flexible techniques for transforming nonlinear issues into high-dimensional linear ones and the dependence on inner products (**Figure 3a**). It's worth noting that the method is inefficient in practice and fails to be implemented in real time especially with large training sets. Kernel methods in feature extraction require the computation of the kernel function between the corresponding sample and all the training ones, hence the efficiency of feature extraction drops with the increase of the number of training samples. Meanwhile, the generalization performance of kernel methods relies on the training set. Thus, these characteristics hinder the wide application of kernel methods.

Decision tree [36] is a mapping between object attributes and values, with leaf nodes representing predicted results to which the object belongs to (**Figure 3b**). The decision tree could be recognized as a set of if-then rules, or conditional probability distributions over the feature and class space. Decision tree is fast, understandable and interpretable. Yet, issues of overfitting and low stability and thus require attention in algorithm selection.

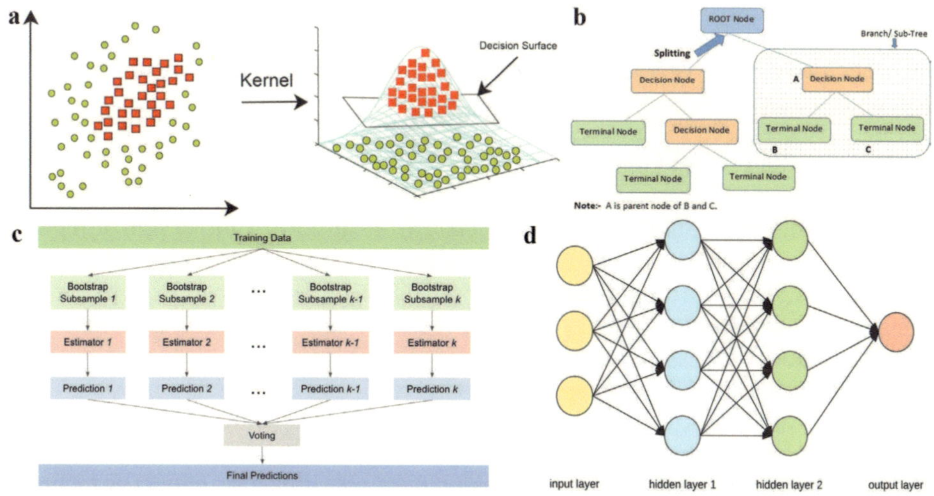

Figure 3.
Schematic diagrams of (a) kernel methods, (b) decision tree [35], (c) ensemble methods, (d) artificial neural network.

Ensemble Methods [37] are mainly carried out by combining multiple learners, which exhibits better generalization performance than individual learner, resulting in strong learners (**Figure 3c**). The error rate of ensemble will decrease exponentially as the number of individual classifiers increases. Gradient Boosting Regression (GBR) [38] is a commonly employed Ensemble Method that sequentially generates multiple weak learners to fit the negative gradient of a previously accumulated model loss function, leading to a gradient decrease in the accumulated model loss gradient after the addition of the weak learners. The combination of multiple weak predictive models can be used to build more powerful predictive models. Random forest [39] is a mainstream ensemble learning method for classification and regression. This method constructs a complete decision tree in parallel by randomly self-sampling the dataset and features. In particular, randomness ensures low correlations between individual trees and reduces bias. Simultaneously, the abundance of trees also reduces overfitting issues.

Deep learning [40] is a kind of ML that realizes the automatic learning and prediction of a large amount of data by imitating the neural network structure of the human brain. Deep learning has many advantages: (i) high accuracy. Comparing to traditional ML methods, deep learning can achieve better results when dealing with large amounts of complex data since features in the data can be automatically extracted. (ii) Highly adaptable. Deep learning models can adapt to different application scenarios by adjusting themselves to different datasets. (iii) Highly scalable. Deep learning models allow to enhance the performance of the model by increasing the number of network layers and nodes to handle large scale data and complex issues.

Artificial neural network (ANN) [41] is the most basic deep learning algorithm derived from the working principles of biological neurons. As shown in **Figure 3d**, patterns can be automatically recognized, classified and predicted by processing and learning the input data, thus simulating the intelligent thinking process of human beings. ANN consists of multiple layers including input layer, hidden layer and output layer. The strength of the connections between neurons is expressed by the weights, and each neuron includes activation function for controlling the activation of the neuron.

Convolutional Neural Network (CNN) [42] is deep Feedforward Neural Networks with the characteristics of local connection and weight sharing. CNN is specialized in processing images such as image classification, target detection, image segmentation and other visual tasks have significant enhancement effect, is currently one of the most widely used models.

2.5 Model evaluation and optimization

Model evaluation and optimization accelerate the discovery and design of high-performance catalysts. At meanwhile, the evaluation model should select appropriate evaluation criteria according to the type of issue. For regression, correlation coefficient (R^2), mean absolute percentage error (MAPE) and root mean square error (RMSE) are selected to evaluate the model. For classification, the evaluation criteria of model accuracy include accuracy, precision, recall, F1 Score, ROC curve, etc. Additionally, model optimization is realized through hyperparameter optimization. For hyperparameter optimization is performed using grid search, random search, Bayesian optimization, etc. And the hyperparameter optimization generally includes the following elements: (i) objective function, which is the objective that the algorithm needs to maximize/minimize, (ii) search range, usually determined by upper and lower limits, and (iii) other parameters of algorithm, such as search step size.

3. ML guides HEAs catalyst discovery

HEAs exhibit excellent catalytic activity for various catalytic activity including electrocatalysis, photocatalysis and enzymatic. HEAs provides a near-continuous distribution in aspects of adsorption energies due to the microstructure (phase structure, electronic structure) and random configuration. A minor number of configurations with optimal properties dominate the overall catalytic activity. We summarized the descriptors and algorithms in ML for catalysis. In this part, the discovery of HEAs catalyst with ML are reported in detail.

3.1 Electrocatalyst

Electrochemical catalysis takes an essential role in energy conversion devices (e.g., fuel cells, proton exchange membrane electrolyzes) [43–45]. The excellent electrocatalytic performance including oxygen reduction reaction (ORR), CO_2 reduction reaction (CO_2RR), hydrogen evolution reaction (HER), N_2 reduction reaction (NRR), etc. are curial to ensure that the power yield of the energy conversation. And the adsorption energies as well as the descriptors are not the identical for different catalytic reactions, thus a specific analysis has to be done for individual reactions. The structure-activity relationship of HEA electrocatalysis is still not comprehensive understanding, and the huge phase space and configuration makes HEA catalysts extraordinary difficult to screen. Meanwhile, ML has the advantage of taking these factors into account and screens catalyst efficiently. Thus, the exploration of efficient HEAs catalysis using ML is essential in energy conversation.

3.1.1 Oxygen reduction reaction (ORR)

ORR acts as a crucial role in sustainable energy systems such as fuel cells and metal–air batteries. And the ORR is a multielectron reaction ($1/2O_2 + 2H^+ + 2e^- = H_2O$) that includes basic steps involving different reaction intermediates. The associative mechanism involves *OOH, *O and *OH intermediates, while the dissociative pathway involves *O and *OH intermediates [46]. Since the adsorption energies of *OH and *OOH scale linearly, it is sufficient to focus on the *OH and O* intermediates to predict the catalytic activity. According to the Sabatier volcano, the optimal binding energy of *OH and *O is 0.1 and 0.2 eV stronger than that of Pt (1 1 1) as suggested by the theory and experiment [47, 48].

Initially, a framework for the IrPdPtRhRu quinary alloy with face-centered cubic (fcc) structure (1 1 1) surface and corresponding ORR activity was built based on DFT calculations and sequential least squares programming (SLSQP) (**Figure 4a**). Consideration of lattice distortion effects, configuration diversity, and atomic ratios of HEAs, based on Sabatier volcano $\left(A = \sum_{i=1}^{z} \left(\prod_{k=1}^{metals} f_k^{n_{ik}} \right) \exp\left(-\frac{\left| \Delta E_i - \Delta E_{opt} \right|}{k_B T} \right) \right)$, is taken to maximize the likelihood of a specific site with the desired adsorption energy. And the optimization of the model yielded an ORR activity and overpotential of 0.10 and 0.04 for $Ir_{10.2}Pd_{32}Pt_{9.3}Rh_{19.6}Ru_{28.9}$ [49]. For the effect of lattice strain, according to the strain correction, the difference in lattice constant between equimolar HEA and optimal HEA is very small and the change in adsorption energy is negligible. Meanwhile, the interaction between ligand and coordination effect on ORR performance of IrPdPtRhRu are explored by the high-fidelity neural network model

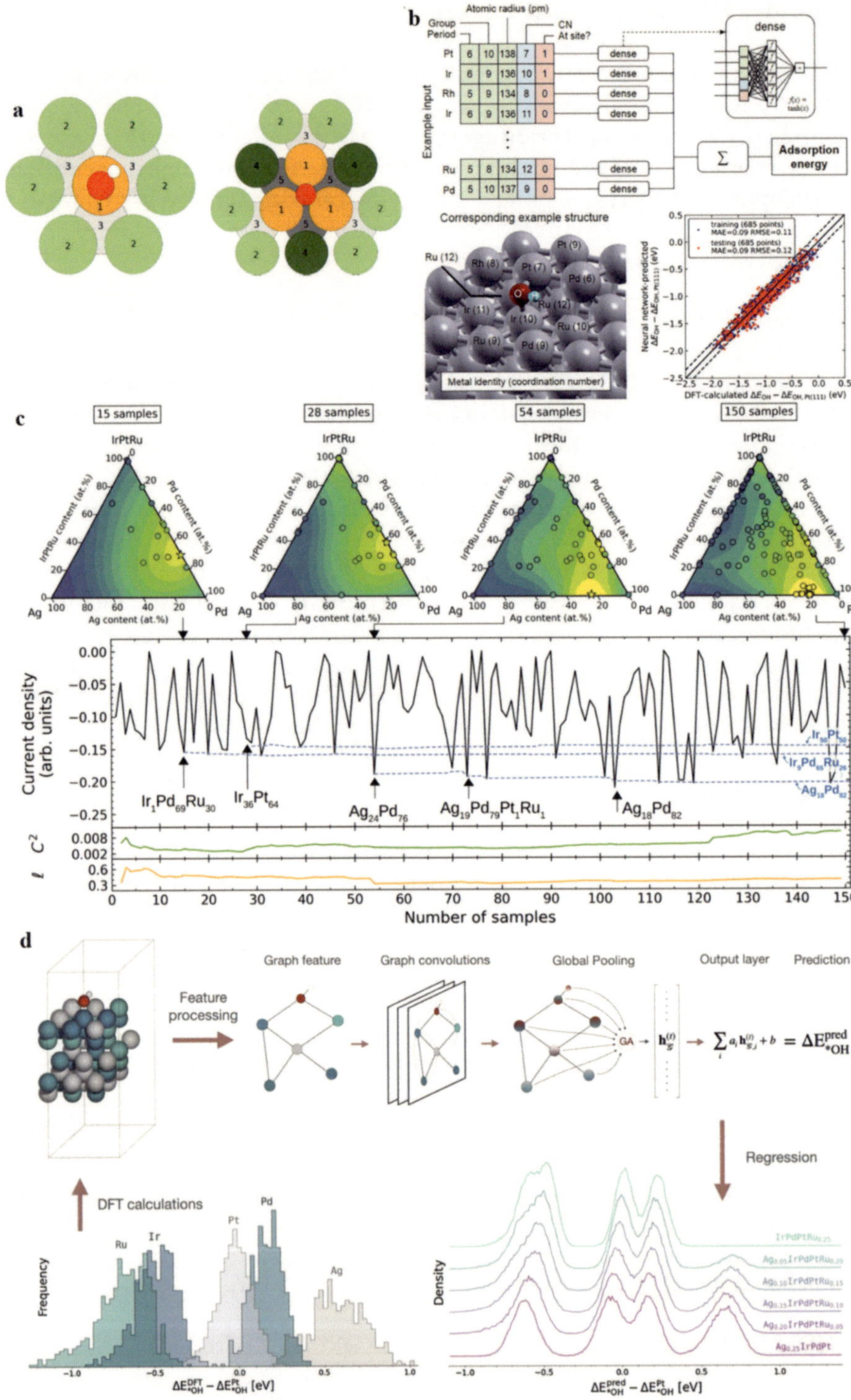

Figure 4.
*(a) Parameters of the surface configurations using nearest neighbors with *OH on-top binding and *O fcc hollow site binding [49]. (b) Machine-learning scheme, example input features, and model parity plot [50]. (c) Example of a Bayesian ORR composition optimization for the quinary Ag-Ir-Pd-Pt-Ru system [51]. (d) Overview of the adsorption energy regression procedure [13].*

(**Figure 4b**) [50]. For the ligand effect on ORR activity, the average adsorption energy follows the order Ru-Ru < Rh-Rh < Ir-Ir < Pd-Pd < Pt-Pt, which is consistent with the tendency in pure metals (Ru < Rh < Ir < Pd < Pt) [52]. And for active sites coordinated with different metals (X-Y), the adsorption energy depends on the average of X-X and Y-Y. For the coordination effect, the means of adsorption energy frequency distributions are linear correlated with the degree of coordination. Due to the numerous configurations, the models obtained for different components differ greatly of HEAs, the vast experiments of HEAs catalysts cannot be accomplished within realistic time scales. Therefore, kinetic model combined with Bayesian optimization is put forward to predict the most active compositions for IrPdPtRhRu and AgIrPdPtRu in ORR (**Figure 4c**) [51]. Relationships are established between the locally optimal component with (1 1 1) crystal surface and the kinetic current density at 0.82 V vs. reversible hydrogen electrode (RHE) through modeling. And the locally optimal compositions are $Ag_{15}Pd_{85}$, $Ir_{50}Pt_{50}$ and $Ir_{10}Pd_{60}Ru_{30}$. It is worth noting that as many of the actual reaction condition parameters are not adequately accounted in model, a direct comparison of experimental data and predicted results is desired in improving the accuracy. While proposing optimal compositions, the model is also capable of estimating the number of experiments required to discover the optimal composition in the large compositional space of HEAs is about 50.

Based on the likelihood of the stable and active HEAs, AgIrPdPtRu has been chosen to investigate for ORR in recent researches [53]. Some sites of a real HEA surface would prefer an *O binding to the hollow position while other sites would prefer an *OH to on-top, and these sites would be blocked by the neighboring adsorption. Upon the surface of a HEA, it is the interactions between the adsorbates and the resulting blockages that produce the active surface. Therefore, a model that incorporates the simplest adsorbate-adsorbate interaction, combining simulation, ML, data-guided combinatorial synthesis and high-throughput characterization, replicates experimental activity trend very well for predicting ORR activity in AgIrPdPtRu [54]. Furthermore, Jan Rossmeisl et al. proposed to a framework to forecast the adsorption energy and couple the net adsorption energy distribution of HEA surface to the expected current density use the graph convolution neural network. And finally, as shown in **Figure 4d**, a serious of optimal compositions of AgIrPdPtRu is obtained by Bayesian optimization [13]. Moreover, based on the AgIrPdPtRu prediction model, catalysts that are already optimal in one compositional space are further optimized in another compositional space. This kind of new strategy for searching the excellent HEA catalyst is efficient in such extensive composition space [55].

In addition, the backward elimination method is utilized to investigate the optimal quinary alloy thought 8-element composition space composed of Au, Ir, Os, Pd, Pt, Re, Rh and Ru, and identify that Au, Pd and Pt have the largest affinity towards the ORR [56]. This kind of strategy would accelerate the screening of high-performance HEA catalysts in large composition space towards ORR.

3.1.2 CO_2 reduction reaction (CO_2RR)

Electrocatalytic CO_2RR is a complex reaction containing multielectron, and the products can be divided into CO, methanol (CH_3OH), formic acid (HCOOH), methane (CH_4) and ethene (CH_2CH_2). Although numerous high efficiency catalysts have been reported, the selectivity and efficiency still cannot satisfy the needs of industrial applications [57]. Therefore, utilizing ML to screen HEAs with superior activity can effectively solve the energy crisis. The binding of Cu to CO was sufficient to inhibit the formation of CO and formic acid, while the binding of Cu to H was weak enough to inhibit the formation of H_2. Thus, Cu plays a crucial role in CO_2RR.

According to the previous studies about the adsorption strengths of CO and H, the CoCuGaNiZn and AgAuCuPdPt HEAs composed of both strong (Co, Ni, Pd and Pt) and weak adsorption strengths (Ag, Au, Ga and Zn) are chosen to search the optimal composition for the performance of CO_2 reduce to CO. The adsorption energies of CO and H with distinct surface microstructures for on-top, fcc-hollow and hcp-hollow are essential input features in accurate prediction (**Figure 5a**) [11]. In the case of CO_2 hydrogenation to methanol, the adsorption energies of H_nCO (n = 0–3) and H are important factors, and the O adsorption is introduced to avoid the oxidation on catalyst surface. Based on the adsorption energies of *CO, *H_2CO and *O on Cu (1 1 1), 35 active catalysts are obtained from the CuCoNiZnSn- and CuCoNiZnMg-based catalysts by supervised learning method, among which 34 are HEAs and 1 is medium entropy alloys [59, 60]. In addition, as for the reaction of CO_2 reduce to HCOOH, the COOH*, CO* and CHO* intermediates are significant parameters. Thus, based on the adsorption energies of COOH*/CHO* and CO* in FeCoNiCuMo HEA and NN ML model, the HEA catalyst with best activity of the lowest potential (0.29 V vs. RHE) is achieved (**Figure 5b**) [58].

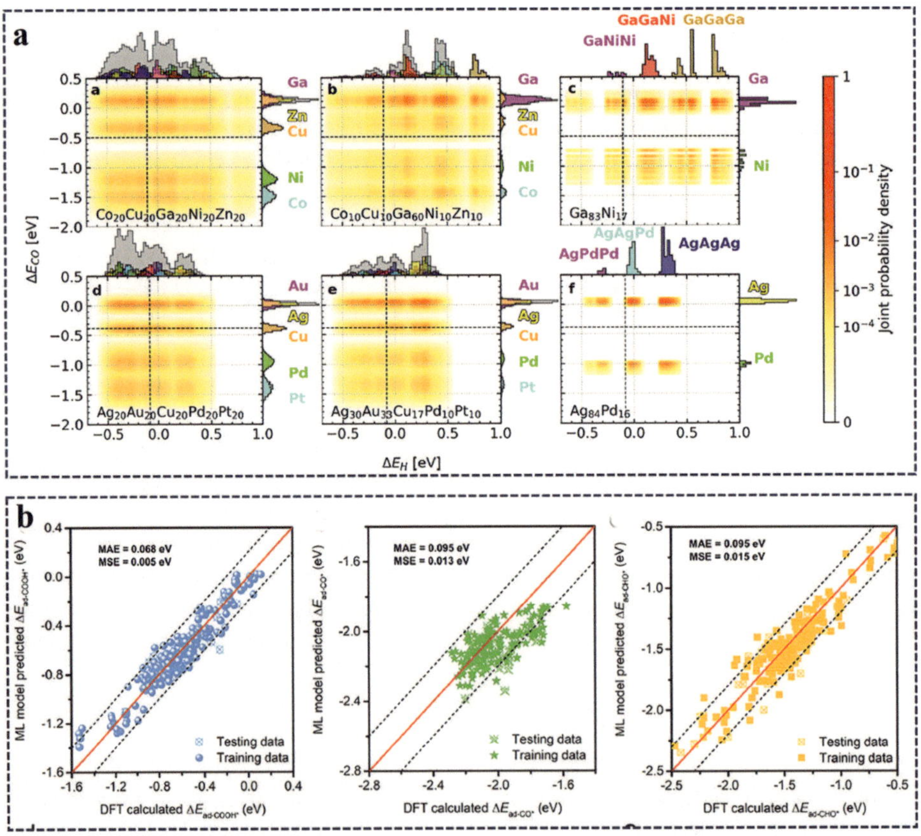

Figure 5.
(a) Adsorption energy distributions [11], (b) NN models predicted the adsorption energies of COOH, CO*, and CHO* plotted against DFT-calculated corresponding adsorption energy (85/15% training–testing data splitting). The dotted lines indicate ±0.20 eV deviation. MAE and MSE values are for testing data [58].*

3.1.3 Hydrogen evolution reaction (HER)

HER is a key reaction in renewable energy such as fuel cell. Up to date, Pt exhibits good performance towards HER due to the proper adsorption energy of H* and low water splitting barrier. Thus, serious Pt-based HEAs catalysts such as PtCoMoPdRh nanoflowers are investigated in recent years in experiment aspect [61–63]. To overcome the problem of high cost and scarcity, noble metal-free HEA catalyst composed of Co, Mo, Fe, Ni and Cu are investigated by ML and hydrogen adsorption energy. Due to the lattice distortion effect of HEAs, the atomic arrangement is disordered and the resulting active sites are continuous. Therefore, the average adsorption energy originated from the numerous different active sites, quantifies the most probable adsorption energy on the surface, is considered as criterion to design CoMoFeNiCu HEA catalyst [64]. The relationship between average adsorption energy and possibilities of a HEA surface microstructure is following: $\overline{\Delta E} = \sum_{\ell}^{n_s} f_\ell \Delta E_\ell / \sum_{\ell}^{n_s} f_\ell$, where $f = \prod_{k}^{4} c_k^{N_k}$ indicates the number of possibilities of a surface microstructure with metal atoms (N_k) in accordance with HEA composition c_k (label ℓ is for clarity). As shown in **Figure 6a**, the HEAs with $\overline{\Delta E}$ = −0.48 eV on average bind *H are expected to be high-performance HER catalysts. Considering the effect of individual constituent metal elements, as shown in **Figure 6b**, it was found that the *H adsorption strength on the surface of HEAs increases with increasing Mo/Fe content and decreases with increasing Cu content.

3.1.4 Nitrogen reduction reaction (NRR)

Ammonia (NH_3) is an effective chemical fertilizer and emerges as potential energy carrier owing to the high hydrogen content and easy liquefaction. Electrocatalysis is an efficient and environmentally friendly method for the conversion of nitrogen to

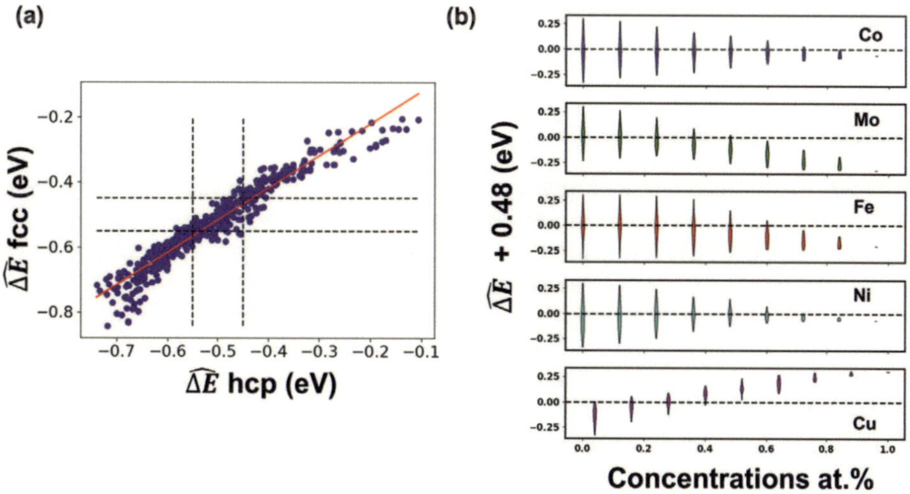

Figure 6.
*(a) Correlations between $\overline{\Delta E}$ for *H at fcc and hcp sites. The red line corresponds to linear dependence with predictivity coefficient r^2 = 0.96. (b) Variations of $\overline{\Delta E}$ with elements compositions. For convenience, we shifted $\overline{\Delta E}$ by −0.48 eV so the zero value indicates catalysts that bind *H optimally as in Pt [64].*

ammonia. The low Faradaic efficiency (FE) and poor stability are the major bottleneck in NRR. Due to the cocktail effect, HEAs show great potential for enhanced the catalytic performance of NRR. RuFeCoNiCu nanoparticles with size of 16 nm exhibit 38.5% FE for NRR in experimental [65]. Aspired by that, Tomas Edvinsson et al. screen over the quinary HEAs in Mo-Cr-Mn-Fe-Co-Ni-Cu-Zn system via DFT calculations and deep neural networks (DNN) for NRR. Considering the average valence electron concentration and electronegativity, the volcano-shaped relationship with $Mo_{0.38}Fe_{0.31}Co_{0.19}Ni_{0.06}Cu_{0.06}$ located on the top of the volcano and $Mo_{0.44}Co_{0.38}Ni_{0.06}Cu_{0.06}Zn_{0.06}$ is an alternative option that balances activity and selectivity [66].

3.1.5 Selective catalytic oxidation of ammonia (NH_3-SCO)

As an energy carrier, NH_3 is emitted up to 40 million tons per year in industry, agriculture, and automobile exhaust. Thus, NH_3-SCO involving multi-intermediates (NH, NH_2, NH_3, etc.) is considered as the effective strategy for depollution. Recent reports demonstrate that the noble metal catalysts such as Pt, Ag, Au and Pd etc. possess promising performance towards NH_3-SCO. In case of the scarcity of noble metals and the cocktail effect of HEAs, DFT-based high-throughput calculation is utilized to screen the adsorption energies of N-atom and O-atom on non-noble metals HEAs. CuZnNiMoCo is selected as the most promising HEA for the NH_3-SCO process due to the combination of constituent metals with strong and weak N-atom binding ability [14]. Meanwhile, according to the volcano curve, the N-atom adsorption energy of the optimal NH_3-SCO activity is $-5.0 \sim -4.0$ eV [67]. Valence electron concentration (VEC) is an important and reliable criterion for recognizing the phase structure of high-entropy alloys. HEAs with VEC ≤ 6.87 and VEC ≥ 8 are corresponding to body-centered cubic (BCC) and FCC structure, respectively [68]. Based on the single-phase HEAs, the probability of the optimal configuration surface containing certain types of atoms is applied to measure the disorder as well as the complexity of the HEA configurations. It's found that CuZnNiMoCo and CuZnNiMoAl have a larger number of optimal configurations than the other CuZn-based HEAs, suggesting that these two non-precious-metal HEAs exhibit a higher NH_3-SCO activity. Furthermore, CuZnNiMoCo and CuZnNiMoAl display a stronger O adsorption ability than Pt (1 1 1), indicating better selectivity in the NH_3-SCO process.

3.1.6 CO oxidation reaction (COOR)

CO will be produced in the process of water gas conversion reaction and methanol fuel cell reaction, and the catalyst with weak anti-CO toxicity performance will lose its inherent activity, resulting in poor stability of the catalyst. Thus, improving the CO tolerance of Pt-based catalysts is crucial in the durability of energy conversion devices. Based on the experiment data, the relationship between the content of constituent metals and the onset potential of COOR was evaluated by Bayesian optimization. The role of constituent metals in onset potential was investigated, and found that Ru possess largest effect on the onset potential of COOR. Combined with DFT theory calculation, in PtRuPdRhAu HEA system, Ru exhibits the largest effect on COOR, which is corresponding to the lowest *OH adsorption energy [69].

3.1.7 Oxygen evolution reaction (OER)

The electrochemical water splitting is composed of OER, ORR and HER. Among them, the OER reaction process involves multiple proton/electron coupling, and more energy barrier need to be overcome, resulting in a sluggish overall reaction rate of water splitting. Therefore, the rational design of high-performance catalysts to enhance OER reactions is a challenge. Lu et al. put forward a cycle of catalyst synthesis-in situ measurement and characterization-database construction-ML analysis-catalyst design to rapid screen the HEAs. The conclusion of the increasing the entropy value improves the electrocatalytic activity is further verified [70].

In addition, the application of HEA catalysts in electrocatalysis is not limited to the reactions discussed above, but also can be applied in methanol oxidation reaction (MOR) [71], glycerol oxidation reaction (GOR) [72], etc. The HEAs would exhibit promising and potential electrochemical activity due to the high-entropy effect, lattice distortion effect, lattice distortion and cocktail effect. However, the large composition space, disordered atomic arrangement and configuration make the screening of catalysts difficult, thus the utilization of ML to accelerate the research of HEA catalysts is conducive to the design of high-performance HEAs and promoting the development of energy conversion.

In the investigation of ML for the application to HEA electrocatalysis, the model is mainly established by calculating the adsorption energies of various configurations of HEAs. In which, the average adsorption energy and probability of optimal HEA surface are introduced into quantize the configurations. Meanwhile, the relationship between electronegativity/entropy value/phase structure and catalytic activity was established. However, there still many microstructures such as electronic structure, charge density and bond length are required to improve the accuracy of prediction. Additionally, the complexity of the experimental conditions, as well as the high-entropy effect and the cocktail effect have a significant impact on the catalytic activity, and thus, the ML model incorporates as many descriptors as possible in order to comprehensively understand the relationship between the structure of the HEAs and the corresponding activity, and to obtain a more highly accurate predictive model that can reasonably guide the discovery of high-performance HEA catalyst.

3.2 Photocatalysis

Photocatalysis, derived by photons, is a reaction that converts solar energy into chemical energy. Photocatalyst absorbs photons whose energy is larger than the band gap of photocatalyst. Electron-hole pairs are generated when electrons shift to the conduction band (CB), resulting holes remain in valence band (VB). Photocatalytic oxides are kind of effective photocatalysts especial in pollutant degradation due to the structure-dependent optical and electronic properties [73]. Inspired by the unique physical and chemical properties of HEAs, high-entropy oxide (HEO) emerges as promising electrode materials and catalysts in the field of Li-ion and Li-S batteries and OER. Thus, the lattice constants and formation energies with variable constituent metals and occupying positions of spinel-structured (Cr, Co, Fe, Mn, and Ni)Ox (HEO_X) were calculated by DFT and predicted by ML (back-propagation network (BPN) and genetic algorithm neural network (GANN)) (**Figure 7**). The occupation of different metal cations in A and B in HEOx affect the lattice constants and formation energies, and the formation energy of HEOx is inversely proportional to the amount of Cr [74].

Additionally, photocatalysts can also be divided into photocatalytic organic and photocatalytic oxides for water splitting. However, owing to the complexity of HEO and the photocatalytic process, few ML modeling investigations of HEO catalysts for water splitting have been reported in the literature. In order to accelerate the discovery of HEO in photocatalytic reactions, it is necessary to comprehensively understand

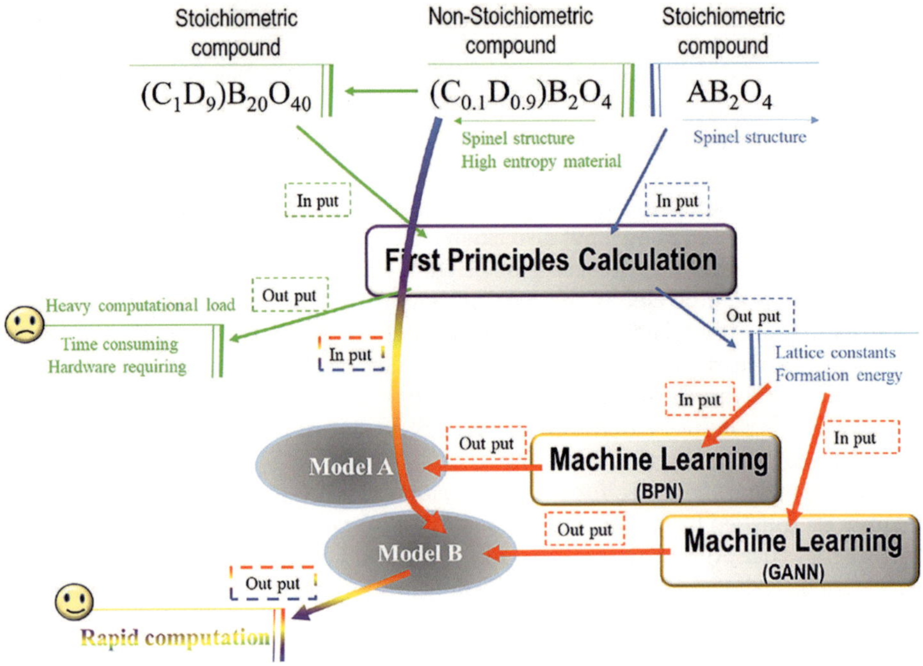

Figure 7.
A schematic of the main concepts of the experiment. The blue region indicates the physical properties of the binary and ternary spinel compounds from the DFT calculations. The green area indicates the non-stoichiometric compound. The red region indicates the calculated physical properties of binary and ternary spinel compounds are used as input files, and then the predicted physical properties of the non-stoichiometric compound are obtained by using ML [74].

the microstructure of HEO and the mechanism of photocatalytic reactions, to construct HEO constructure-activity relationships, to introduce sufficient feature, and to improve the accuracy of the model.

4. Conclusion

In this chapter, we summarize the application of ML in HEA/HEO catalysts, the commonly used ML algorithms and the features for electro–/photo-catalysts. And the relationship between HEA microstructure (configurations, intermediate adsorption energy, electronegativity, etc.) and catalytic activity is established. The HEA catalysts would make promising to the other field of catalysis such as enzyme catalysis, photo-electrocatalysis and thermo-catalysis. However, the high-entropy effect and the cocktail effect in aspect of atomic and electronic structure of HEA are insufficiently explored. Meanwhile, And Complexity under actual operating conditions essentially affect the accuracy of ML model. Thus, in order to acquire the accusative prediction model and high-performance HEA catalysts, there are two strategies: (i) based on the current optimized algorithm, fully considering the complexity of the experimental environment is a necessary strategy to improve the accuracy. (ii) With the universalization of machine learning, choosing the appropriate algorithm to establish the model effectively minimizes the volume of DFT calculations as well as obtains more accurate prediction.

Conflict of interest

The authors declare no conflict of interest.

Author details

Jike Wang*, Min Wei and Junyu Zhang
The Institute for Advanced Studies, Wuhan University, Wuhan, China

*Address all correspondence to: jike.wang@whu.edu.cn

References

[1] Wang Y, Wang Y. High-entropy alloys in catalyses and supercapacitors: Progress, prospects. Nano Energy. 2022;**104**:107958. DOI: 10.1016/j.nanoen.2022.107958

[2] Yao Y, Dong Q, Brozena A, Luo J, Miao J, Chi M, et al. High-entropy nanoparticles: Synthesis-structure-property relationships and data-driven discovery. Science. 2022;**376**(6589):eabn3103. DOI: 10.1126/science.abn3103

[3] Sun Y, Dai S. High-entropy materials for catalysis: A new frontier. Science Advances. 2021;7(20):2375-2548. DOI: 10.1126/sciadv.abg1600

[4] George EP, Raabe D, Ritchie RO. High-entropy alloys. Nature Reviews Materials. 2019;**4**(8):515-534. DOI: 10.1038/s41578-019-0121-4

[5] Yeh JW, Chen SK, Lin SJ, Gan JY, Chin TS, Shun TT, et al. Nanostructured high-entropy alloys with multiple principal elements: Novel alloy design concepts and outcomes. Advanced Engineering Materials. 2004;**6**(5):299-303. DOI: 10.1002/adem.200300567

[6] Xin Y, Li S, Qian Y, Zhu W, Yuan H, Jiang P, et al. High-entropy alloys as a platform for catalysis: Progress, challenges, and opportunities. ACS Catalysis. 2020;**10**(19):11280-11306. DOI: 10.1021/acscatal.0c03617

[7] Miracle D, Miller J, Senkov O, Woodward C, Uchic M, Tiley J. Exploration and development of high entropy alloys for structural applications. Entropy. 2014;**16**(1):494-525. DOI: 10.3390/e16010494

[8] Sarker P, Harrington T, Toher C, Oses C, Samiee M, Maria J-P, et al. High-entropy high-hardness metal carbides discovered by entropy descriptors. Nature Communications. 2018;**9**(1):4980. DOI: 10.1038/s41467-018-07160-7

[9] Zhai Y, Ren X, Wang B, Liu S. High-entropy catalyst—A novel platform for electrochemical water splitting. Advanced Functional Materials. 2022;**32**(47):2207536. DOI: 10.1002/adfm.202207536

[10] Gorsse S, Nguyen MH, Senkov ON, Miracle DB. Database on the mechanical properties of high entropy alloys and complex concentrated alloys. Data in Brief. 2018;**21**:2664-2678. DOI: 10.1016/j.dib.2018.11.111

[11] Pedersen JK, Batchelor TAA, Bagger A, Rossmeisl J. High-entropy alloys as catalysts for the CO_2 and CO reduction reactions. ACS Catalysis. 2020;**10**(3):2169-2176. DOI: 10.1021/acscatal.9b04343

[12] Yuan G, Wu M, Ruiz PL. Density functional theory-machine learning characterization of the adsorption energy of oxygen intermediates on high-entropy alloys made of earth-abundant metals. The Journal of Physical Chemistry C. 2023;**127**(32):15809-15818. DOI: 10.1021/acs.jpcc.3c03404

[13] Clausen CM, Nielsen MLS, Pedersen JK, Rossmeisl J. Ab initio to activity: Machine learning-assisted optimization of high-entropy alloy catalytic activity. High Entropy Alloys & Materials. 2022;**1**:120-133. DOI: 10.1007/s44210-022-00006-4

[14] Chen L, Yang J, Zhang Z, Wu F, Yao Y, Zhang A, et al. High throughput screening of Noble metal-free

high-entropy alloys catalysts for selective catalytic oxidation of NH_3. Applied Surface Science. 2023;**628**:157354. DOI: 10.1016/j.apsusc.2023.157354

[15] Himanen L, Jäger MOJ, Morooka EV, Federici Canova F, Ranawat YS, Gao DZ, et al. Dscribe: Library of descriptors for machine learning in materials science. Computer Physics Communications. 2020;**247**:106949. DOI: 10.1016/j.cpc.2019.106949

[16] Sun M, Dougherty AW, Huang B, Li Y, Yan CH. Accelerating atomic catalyst discovery by theoretical calculations-machine learning strategy. Advanced Energy Materials. 2020;**10**(12):1903949. DOI: 10.1002/aenm.201903949

[17] Parker AJ, Opletal G, Barnard AS. Classification of platinum nanoparticle catalysts using machine learning. Journal of Applied Physics. 2020;**128**(1):014301. DOI: 10.1063/5.0009129

[18] Li H, Xu S, Wang M, Chen Z, Ji F, Cheng K, et al. Computational design of (100) alloy surfaces for the hydrogen evolution reaction. Journal of Materials Chemistry A. 2020;**8**(35):17987-17997. DOI: 10.1039/d0ta04615a

[19] Wu L, Guo T, Li T. Machine learning-accelerated prediction of overpotential of oxygen evolution reaction of single-atom catalysts. iScience. 2021;**24**(5):102398. DOI: 10.1016/j.isci.2021.102398

[20] Wu L, Guo T, Li T. Rational Design of Transition Metal Single-Atom Electrocatalysts: A simulation-based, machine learning-accelerated study. Journal of Materials Chemistry A. 2020;**8**(37):19290-19299. DOI: 10.1039/d0ta06207c

[21] Esterhuizen JA, Goldsmith BR, Linic S. Theory-guided machine learning finds geometric structure-property relationships for chemisorption on subsurface alloys. Chem. 2020;**6**(11):3100-3117. DOI: 10.1016/j.chempr.2020.09.001

[22] Zheng J, Sun X, Qiu C, Yan Y, Yao Z, Deng S, et al. High-throughput screening of hydrogen evolution reaction catalysts in Mxene materials. The Journal of Physical Chemistry C. 2020;**124**(25):13695-13705. DOI: 10.1021/acs.jpcc.0c02265

[23] Ge L, Yuan H, Min Y, Li L, Chen S, Xu L, et al. Predicted optimal bifunctional electrocatalysts for the hydrogen evolution reaction and the oxygen evolution reaction using chalcogenide heterostructures based on machine learning analysis of in silico quantum mechanics based high throughput screening. Journal of Physical Chemistry Letters. 2020;**11**(3):869-876. DOI: 10.1021/acs.jpclett.9b03875

[24] Wexler RB, Martirez JMP, Rappe AM. Chemical pressure-driven enhancement of the hydrogen evolving activity of Ni_2P from nonmetal surface doping interpreted via machine learning. Journal of the American Chemical Society. 2018;**140**(13):4678-4683. DOI: 10.1021/jacs.8b00947

[25] Gao W, Chen Y, Li B, Liu SP, Liu X, Jiang Q. Determining the adsorption energies of small molecules with the intrinsic properties of adsorbates and substrates. Nature Communications. 2020;**11**(1):1196. DOI: 10.1038/s41467-020-14969-8

[26] Yang Z, Gao W, Jiang Q. A machine learning scheme for the catalytic activity of alloys with intrinsic descriptors. Journal of Materials Chemistry A. 2020;**8**(34):17507-17515. DOI: 10.1039/d0ta06203k

[27] Ran N, Sun B, Qiu W, Song E, Chen T, Liu J. Identifying metallic

transition-metal dichalcogenides for hydrogen evolution through multilevel high-throughput calculations and machine learning. Journal of Physical Chemistry Letters. 2021;**12**(8):2102-2111. DOI: 10.1021/acs.jpclett.0c03839

[28] Guo X, Lin S, Gu J, Zhang S, Chen Z, Huang S. Simultaneously achieving high activity and selectivity toward two-Electron O_2 electroreduction: The power of single-atom catalysts. ACS Catalysis. 2019;**9**(12):11042-11054. DOI: 10.1021/acscatal.9b02778

[29] Maulud D, Abdulazeez AM. A review on linear regression comprehensive in machine learning. Journal of Applied Science and Technology Trends. 2020;**1**(4):140-147. DOI: 10.38094/jastt1457

[30] Marquardt DW. An algorithm for least-squares estimation of nonlinear parameters. Journal of the Society for Industrial and Applied Mathematics. 1963;**11**(2):431-441. DOI: 10.1137/0111030

[31] Hofmann T, Schölkopf B, Smola AJ. Kernel methods in machine learning. The Annals of Statistics. 2008;**36**(3):1171-1220. DOI: 10.1214/009053607000000677

[32] Hearst MA, Dumais ST, Osuna E, Platt J, Scholkopf B. Support vector machines. IEEE Intelligent Systems and Their Applications. 1998;**13**(4):18-28. DOI: 10.1109/5254.708428

[33] Williams C, Rasmussen C. Gaussian processes for regression. Advances in Neural Information Processing Systems. 1995;**8**:515-520. DOI: 10.5555/2998828.2998901

[34] An S, Liu W, Venkatesh S. Fast cross-validation algorithms for least squares support vector machine and kernel ridge regression. Pattern Recognition. 2007;**40**(8):2154-2162. DOI: 10.1016/j.patcog.2006.12.015

[35] Sneha N, Gangil T. Analysis of diabetes mellitus for early prediction using optimal features selection. Journal of Big Data. 2019;**6**(1):1-19. DOI: 10.1186/s40537-019-0175-6

[36] Song YY, Lu Y. Decision tree methods: Applications for classification and prediction. Shanghai Archives of Psychiatry. 2015;**27**(2):130-135. DOI: 10.11919/j.issn.1002-0829.215044

[37] Dietterich TG. Ensemble methods in machine learning. In: International Workshop on Multiple Classifier Systems (MCS 2000), 21-23 June 2000, Cagliari, Italy. Berlin, Heidelberg: Springer; 2000. pp. 1-15

[38] Friedman JH. Greedy function approximation: A gradient boosting machine. The Annals of Statistics. 2001;**29**(5):1189-1232. DOI: 10.1214/aos/1013203451

[39] Breiman L. Random forests. Machine Learning. 2001;**45**(1):5-32. DOI: 10.1023/a:1010933404324

[40] LeCun Y, Bengio Y, Hinton G. Deep learning. Nature. 2015;**521**(7553):436-444. DOI: 10.1038/nature14539

[41] Jain AK, Jianchang M, Mohiuddin KM. Artificial neural networks: A tutorial. Computer. 1996;**29**(3):31-44. DOI: 10.1109/2.485891

[42] Albawi S, Mohammed TA, Al-Zawi S. Understanding of a convolutional neural network. In: International Conference on Engineering and Technology (ICET'17), 21-23 August 2017, Antalya, Turkey. New York: IEEE; 2018. pp. 1-6

[43] Chen G, Wang T, Zhang J, Liu P, Sun H, Zhuang X, et al. Accelerated hydrogen evolution kinetics on

Nife-layered double hydroxide electrocatalysts by tailoring water dissociation active sites. Advanced Materials. 2018;**30**(10):1706279. DOI: 10.1002/adma.201706279

[44] Kodama K, Nagai T, Kuwaki A, Jinnouchi R, Morimoto Y. Challenges in applying highly active Pt-based nanostructured catalysts for oxygen reduction reactions to fuel cell vehicles. Nature Nanotechnology. 2021;**16**(2):140-147. DOI: 10.1038/s41565-020-00824-w

[45] Handoko AD, Wei F, Jenndy YBS, Seh ZW. Understanding heterogeneous electrocatalytic carbon dioxide reduction through operando techniques. Nature Catalysis. 2018;**1**(12):922-934. DOI: 10.1038/s41929-018-0182-6

[46] Kulkarni A, Siahrostami S, Patel A, Nørskov JK. Understanding catalytic activity trends in the oxygen reduction reaction. Chemical Reviews. 2018;**118**(5):2302-2312. DOI: 10.1021/acs.chemrev.7b00488

[47] Stephens IEL, Bondarenko AS, Grønbjerg U, Rossmeisl J, Chorkendorff I. Understanding the electrocatalysis of oxygen reduction on platinum and its alloys. Energy & Environmental Science. 2012;**5**(5):6744-6762. DOI: 10.1039/c2ee03590a

[48] Greeley J, Stephens IEL, Bondarenko AS, Johansson TP, Hansen HA, Jaramillo TF, et al. Alloys of platinum and early transition metals as oxygen reduction electrocatalysts. Nature Chemistry. 2009;**1**(7):552-556. DOI: 10.1038/nchem.367

[49] Batchelor TAA, Pedersen JK, Winther SH, Castelli IE, Jacobsen KW, Rossmeisl J. High-entropy alloys as a discovery platform for electrocatalysis. Joule. 2019;**3**(3):834-845. DOI: 10.1016/j.joule.2018.12.015

[50] Lu Z, Chen ZW, Singh CV. Neural network-assisted development of high-entropy alloy catalysts: Decoupling ligand and coordination effects. Matter. 2020;**3**(4):1318-1333. DOI: 10.1016/j.matt.2020.07.029

[51] Pedersen JK, Clausen CM, Krysiak OA, Xiao B, Batchelor TAA, Löffler T, et al. Bayesian optimization of high-entropy alloy compositions for electrocatalytic oxygen reduction. Angewandte Chemie International Edition. 2021;**60**(45):24144-24152. DOI: 10.1002/anie.202108116

[52] Nørskov JK, Rossmeisl J, Logadottir A, Lindqvist L, Kitchin JR, Bligaard T, et al. Origin of the overpotential for oxygen reduction at a fuel-cell cathode. Journal of Physical Chemistry B. 2004;**108**(46):17886-17892. DOI: 10.1021/jp047349j

[53] Wan X, Zhang Z, Yu W, Niu H, Wang X, Guo Y. Machine-learning-assisted discovery of highly efficient high-entropy alloy catalysts for the oxygen reduction reaction. Patterns (N Y). 2022;**3**(9):100553. DOI: 10.1016/j.patter.2022.100553

[54] Batchelor TAA, Löffler T, Xiao B, Krysiak OA, Strotkötter V, Pedersen JK, et al. Complex-solid-solution electrocatalyst discovery by computational prediction and high-throughput experimentation**. Angewandte Chemie International Edition. 2021;**60**(13):6932-6937. DOI: 10.1002/anie.202014374

[55] Plenge MK, Pedersen JK, Mints VA, Arenz M, Rossmeisl J. Following paths of maximum catalytic activity in the composition space of high-entropy alloys. Advanced Energy Materials. 2022;**13**(2):2202962. DOI: 10.1002/aenm.202202962

[56] Mints VA, Pedersen JK, Wiberg GKH, Rossmeisl J, Arenz M. Backward elimination: A strategy for high-entropy alloy catalyst discovery. ChemRxiv. 2022. DOI: 10.26434/chemrxiv-2022-78s83

[57] Roy D, Das A, Manna S, Pathak B. A route map of machine learning approaches in heterogeneous Co2 reduction reaction. The Journal of Physical Chemistry C. 2023;**127**(2):871-881. DOI: 10.1021/acs.jpcc.2c06924

[58] Chen ZW, Gariepy Z, Chen L, Yao X, Anand A, Liu S-J, et al. Machine-learning-driven high-entropy alloy catalyst discovery to circumvent the scaling relation for CO_2 reduction reaction. ACS Catalysis. 2022;**12**(24):14864-14871. DOI: 10.1021/acscatal.2c03675

[59] Roy D, Mandal SC, Pathak B. Machine learning assisted exploration of high entropy alloy-based catalysts for selective CO_2 reduction to methanol. Journal of Physical Chemistry Letters. 2022;**13**(25):5991-6002. DOI: 10.1021/acs.jpclett.2c00929

[60] Roy D, Mandal SC, Pathak B. Machine learning-driven high-throughput screening of alloy-based catalysts for selective CO_2 hydrogenation to methanol. ACS Applied Materials & Interfaces. 2021;**13**(47):56151-56163. DOI: 10.1021/acsami.1c16696

[61] Wei M, Sun Y, Ai F, Xi S, Zhang J, Wang J. Stretchable high-entropy alloy nanoflowers enable enhanced alkaline hydrogen evolution catalysis. Applied Catalysis B: Environmental. 2023;**334**(5):12814. DOI: 10.1016/j.apcatb.2023.122814

[62] Gao S, Hao S, Huang Z, Yuan Y, Han S, Lei L, et al. Synthesis of high-entropy alloy nanoparticles on supports by the fast moving bed pyrolysis. Nature Communications. 2020;**11**(1):2016. DOI: 10.1038/s41467-020-15934-1

[63] Feng G, Ning F, Song J, Shang H, Zhang K, Ding Z, et al. Sub-2 nm ultrasmall high-entropy alloy nanoparticles for extremely superior electrocatalytic hydrogen evolution. Journal of the American Chemical Society. 2021;**143**(41):17117-17127. DOI: 10.1021/jacs.1c07643

[64] Saidi WA, Nandi T, Yang T. Designing multinary Noble metal-free catalyst for hydrogen evolution reaction. Electrochemical Science Advances. 2023;**3**(6):e2100224. DOI: 10.1002/elsa.202100224

[65] Zhang D, Zhao H, Wu X, Deng Y, Wang Z, Han Y, et al. Multi-site electrocatalysts boost pH-universal nitrogen reduction by high-entropy alloys. Advanced Functional Materials. 2020;**31**(9):2006939. DOI: 10.1002/adfm.202006939

[66] Araujo RB, Bayrak Pehlivan I, Edvinsson T. High-entropy alloy catalysts: Fundamental aspects, promises towards electrochemical NH_3 production, and lessons to learn from deep neural networks. Nano Energy. 2023;**105**:108027. DOI: 10.1016/j.nanoen.2022.108027

[67] Yang J, Chen X, Liu Z, Wang Q, Wen Y, Zhang A, et al. Theoretical design principles of metal catalysts for selective Ammonia oxidation from high throughput computation. Journal of Materials Chemistry A. 2022;**10**(23):12447-12457. DOI: 10.1039/d2ta02326a

[68] Yang X, Zhang Y. Prediction of high-entropy stabilized solid-solution in multi-component alloys. Materials Chemistry and Physics. 2012;**132**(2-3):233-238. DOI: 10.1016/j.matchemphys.2011.11.021

[69] Mints V, Pedersen J, Bagger A, Quinson J, Rossmeisl J, Arenz M. Exploring the composition space of high-entropy alloy nanoparticles for the electrocatalytic H_2/CO oxidation with bayesian optimization. ChemRxiv. 2022;**12**(18):11263-11271. DOI: 10.26434/chemrxiv-2021-zpbqb

[70] Liu C, Ding Y, Guan Y, Tang J, Jiang C, Gao H, et al. Combination of rapid intrinsic activity measurements and machine learning as a screening approach for multicomponent electrocatalysts. ACS Applied Materials & Interfaces. 2023;**15**(36):42532-42540. DOI: 10.1021/acsami.3c07442

[71] Subhash B, Unocic R, Gallington L, Wright J, Cheong S, Tilley R, et al. Resolving atomic-scale structure and chemical coordination in high entropy alloy electrocatalysts for structure-function relationship elucidation. ACS Nano. 2022;**17**(22):22299-22312. DOI: 10.21203/rs.3.rs-2155469/v1

[72] Fan L, Ji Y, Wang G, Chen J, Chen K, Liu X, et al. High entropy alloy electrocatalytic electrode toward alkaline glycerol valorization coupling with acidic hydrogen production. Journal of the American Chemical Society. 2022;**144**(16):7224-7235. DOI: 10.1021/jacs.1c13740

[73] Mai H, Chen D, Tachibana Y, Suzuki H, Abe R, Caruso RA. Developing sustainable, high-performance perovskites in photocatalysis: Design strategies and applications. Chemical Society Reviews. 2021;**50**(24):13692-13729. DOI: 10.1039/d1cs00684c

[74] Lin C-C, Chang C-W, Kaun C-C, Su Y-H. Stepwise evolution of photocatalytic spinel-structured $(Co,Cr,Fe,Mn,Ni)_3O_4$ high entropy oxides from first-principles calculations to machine learning. Crystals. 2021;**11**(9):1035. DOI: 10.3390/cryst11091035

Chapter 4

Molecular Dynamics on Hf-Nb-Ta-Ti-Zr High Entropy Alloy

Luis César R. Aliaga, Alexandre Melhorance Barboza, Loena Marins de Couto and Ivan Napoleão Bastos

Abstract

Classical molecular dynamics simulations were used to investigate the structure and mechanical properties in the equiatomic Hf-Nb-Ta-Ti-Zr high entropy alloy. The open-source code LAMMPS was used to generate alloys with different crystalline lattices to determine the stable structure at 300 K. Alloying elements interacted under the action of the MEAM interatomic potential. The result showed that the alloy stabilizes in body-centered cubic (BCC) structure at 300 K. However, a wide dispersion of potential energy data as a function of atomic separation suggests the coexistence of another crystalline phase. Heating tests indicated a polymorphic phase transformation from BCC to hexagonal close-packed (HCP) at temperatures around 1100 K. Uniaxial tensile tests at a rate of 1×10^{10} s^{-1} along the [001], [110], and [111] crystallographic directions in cylindrical monocrystalline bars at 300 K were conducted. The results revealed a strong anisotropy of mechanical properties. This work provides a microscopic understanding of the mechanical behavior of the multicomponent alloy and aligns with the macroscopic theory of plastic deformation of single crystals.

Keywords: high entropy alloys, molecular dynamics, computer simulations, solid solutions, phase transformations, mechanical properties

1. Introduction

For a long period, the development of technological alloys has been grounded on the traditional concept of solvents and solutes. The main goal of this approach was to improve the properties of the solvent element, also known as matrix, by incorporating certain quantities of solutes [1, 2]. In various situations, this addition can lead to the formation of a second phase that modifies several properties. However, this paradigm changed in 2004 when the concept of high entropy alloys (HEAs) was postulated as single-phase alloys [3, 4], where all the elements have similar importance. HEAs are based on the metallurgical concept that several chemical elements form mainly a disordered solid solution phase [5]. The content of each multiple principal element ranges from 5 to 35 atom percent [6, 7]. Due to its chemical homogeneity, HEAs crystallized mainly in one of the simple metallic lattice structures, such as face-centered cubic (FCC), body-centered cubic (BCC), or hexagonal close-packed (HCP). HEAs present a set of superior mechanical properties such as high ductility, toughness, creep, hardness and fatigue resistance when compared to conventional metallic alloys [8]. This outstanding array of physical and mechanical properties [9] opens up possibilities for applications in various fields such as structures [10], refractory [11], cryogenic [12], energy storage [13], biomedical [14], aerospace [15], electrocatalysis [16], among others [17].

Based on the concept of entropy, researchers have identified two types of alloy systems [18]. The first type is named as HEAs due to its high configurational entropy. Usually, these alloys are composed by at least five elements mixed in equiatomic amounts [19], or in compositions close to that. Due to their high configurational entropy of mixing, the most probable microstructure is composed of a single stabilized solid solution phase [20]. However, in some cases, especially when compositions are out of the equiatomic fraction, HEAs do not exhibit single-phase microstructure [21]. Moreover, HEAs can be formed in relatively large compositional fractions. The second type of alloy system is termed as medium entropy alloys (MEAs) [22]. In this case, the alloys are composed by three or four multi-principal elements and similarly to the HEAs, the MEAs show improved mechanical and physical properties in comparison with ordinary alloys. Although the criterion based on the number of multielements is valid for most compositions, there are exceptions to this classification [23].

While HEAs or multi-principal element alloys have been designed with a focus on improving mechanical and chemical properties, it is notable that among the extensive array of alloy systems, the development of lightweight alloys is crucial for industries such as aeronautics and transportation. Furthermore, because over the years various HEAs have been developed, different reviews have been written paying attention to the processing routes [24], microstructure [25], mechanical [9, 26], and chemical properties [27]. Nonetheless, during the last years, diverse computer simulation techniques have been used to study and design new entropic alloys [28, 29]. Even for equiatomic composition, the number of combinations of alloys with four or five elements, considering all metallic elements of periodic table, is enormous. Therefore, the use of previous computational modeling is crucial in the design of entropic alloys.

Refractory HEAs have been identified as promising candidate materials for aerospace propulsion systems, land-based gas turbines, nuclear reactors, heat exchanger tubing [30], and other applications in the chemical process industry. Among those systems, the equiatomic Hf-Nb-Ta-Ti-Zr HEA stands out by possessing outstandingly high strength combined with good ductility at room temperature [31]. Also, that alloy has attracted attention in the fields of hydrogen storage [32].

In this sense, the present work aims to present the application of molecular dynamics simulation on the study of structure and mechanical behavior at room temperature of the equiatomic Hf-Nb-Ta-Ti-Zr HEA due to its several possibilities of applications.

1.1 Classical molecular dynamics

Several semi-empirical methodologies developed for predicting HEAs provide partial success, and their predictions are not entirely satisfactory. Although the resulting parameters offer valuable insights into the phase formation of HEAs, it is not possible to generalize to all alloy systems and find a comprehensive model remains as a significant challenge to the scientific community.

On the other hand, with the advances in the development of powerful computer devices and robust algorithms, various tools now exist to assist the search for the most promising HEA compositions, whether they exhibit a single or multiphase structure. Among those different approaches, we can find the calculation of phase diagrams (CALPHAD) [33], ab-initio molecular dynamics (AIMD) simulation [34], Monte-Carlo (MC) simulations [35], classical molecular dynamics (MD) [36–38] simulations, and the more recently developed machine learning approach [39–41]. Each approach has its own advantages compared to experimental research. Notably, cost savings can be achieved as all experiments are conducted within a computer, which is more economical than traditional experimental setups. Nevertheless, it is important to highlight that to obtain satisfactory results is necessary to have access to extensive base data [42, 43] or, regardless the study type, to count with powerful computer capacity.

The MD methodology allows for determining both the equilibria as well as out-of-equilibrium thermodynamic properties. In addition, the dynamical properties of a system at finite temperature are also simulated by MD. MD is an efficient tool for exploring the type of structure, thermodynamics, physical, and mechanical properties, offering a nanoscale understanding of the structure and deformation behaviors of materials. However, the quality of MD simulations largely depends on the method used to specify forces, specifically, the interatomic potential among the atoms or force field function. Thus, the interatomic potential is considered as the heart of the simulations, and the accuracy of the obtained results strongly depends on it.

It is widely recognized that there are currently numerous open-source as well as commercial packages available for conducting MD simulations. A simple digital search on MD open-source packages reveals various options each one with its own characteristics. Nonetheless, it is worth noting that most of scientific papers in the field of materials science use the large-scale atomic/molecular massively parallel simulator (LAMMPS) package [44, 45].

LAMMPS is a flexible simulation tool for particle-based materials modeling at the atomic, meso, and continuum scales, which can simulate both the MD and MC approaches. Additionally, it is an open-source code that operates on various operational systems, including Windows, Linux, or Mac. Furthermore, although LAMMPS does not have a user-friendly interface its programming is not too complicated. This tool is accessible for both trained programmers and novice researchers.

1.1.1 Interatomic potential

Molecular dynamics simulation is divided into two categories. The first category corresponds to AIMD or first principles simulation, and the second category corresponds to the classical MD. AIMD is a powerful tool with a great accuracy in predicting the parameters and behaviors of different materials. Its principal drawback is the significant amount of time required to obtain the results, in addition to supporting a small number of atoms. Nonetheless, it is strongly recommended to study the physical and chemical properties of materials. On the other hand, classical MD is faster than AIMD, although at the expense of a slight reduction in prediction accuracy. MD is based on Newton's second law, where its primary control mechanism is the force field or interatomic potential (IP) that governs the interaction of atoms. The IP plays a fundamental role in the MD simulations, and the predictive accuracy depends on how well it describes the interaction among atoms.

All the IPs enable the rapid calculations of the total energy of the atomic systems, facilitating the behavior simulation of metals, ceramics, polymers, and composites. The development of any IP should consider a compromise between computational speed and the accuracy of modeling. For each type of material, it is necessary to consider specific aspects to adequately represent atomic bonding. Hence, for instance, an IP for ceramics should differ from that of the metallic alloy.

IPs also known as force fields, are mathematical functions dependent on atomic coordinates, approximating either the electronic ground state energy or, in the case of metals, the electronic free energy within a system [46]. In MD simulations, the forces acting on atoms, essential for their motion, can be determined by computing the gradient of these functions with respect to the atomic coordinates. Understanding the importance of IP is essential for accurately predicting and interpreting the dynamic evolution of a molecular system.

There are several types of IPs. However, here we are focusing just on the IP used to simulate the equiatomic HEA that corresponds to one embedded atom method (EAM) type. Daw and Baskes [47, 48] proposed an alternative to the pair potential description by employing density functional ideas. This approach, known as the EAM potential, has been applied in MD to characterize diverse and intricate physical scenarios in metallic systems [49]. EAM potentials distinguish themselves from two-body potentials by incorporating a many-body term, dependent on an effective electron density, alongside the pair potential. This many-body term is employed to encapsulate the metallic bonding of the system [50].

In the EAM, each atom is considered to be embedded within the surrounding electron density contributed by neighboring atoms. As a result, the potential energy of a group of atoms is determined by the sum of the pair interaction energy between nuclei of atoms i and j and the embedding energy. The embedding energy is a function of the local background electron density around the i-th atom [51]. The EAM is particularly well-suited for investigating systems featuring crystalline defects such as dislocations and grain boundaries, owing to its capacity to handle local background electron density. However, a crucial assumption in the derivation of the EAM expression is that the electronic cloud around each atom has a spherical shape. While this approximation holds true for FCC crystal structures, the EAM faces challenges in accurately representing systems where directional bonding is crucial, such as BCC, HCP, and nonmetallic materials [51]. The total energy of a system with N atoms within the EAM formalism is given by:

$$U_{\mathrm{EAM}} = \sum_{i=1}^{N} \left[F_i(\rho_i) + \frac{1}{2} \sum_{\substack{j=1 \\ j\neq i}}^{N} u_{ij}\left(r_{ij}\right) \right] \tag{1}$$

where $F_i(\rho_i)$ is the energy required to insert atom i into the electron density ρ_i at position i, and $u_{ij}\left(r_{ij}\right)$ is the pair-interaction potential between atom i and j with distance r_{ij}. The electron density can be expressed as:

$$\rho_i = \sum_{\substack{j=1 \\ j\neq i}}^{N} \phi\left(r_{ij}\right) \tag{2}$$

where $\phi(r_{ij})$ is the electron density of atom j at a distance r_{ij} from the nucleus of atom i. Used either independently or in a hybrid configuration, the EAM potential is extensively applied in the study of HEA [30, 52–55].

EAM potentials demonstrate a limitation in accurately representing non-close-packed structures, specifically those differing from FCC. To enable a single formalism for a broad range of structures and elements, including FCC, BCC, HCP, diamond-structured, and even gaseous elements, Baskes [56] developed the modified embedded-atom method (MEAM). Similar to the EAM, the MEAM formulation includes pairwise repulsions and an embedding function. However, unlike the EAM, the MEAM potential introduces angular-dependent interactions through the electron density term. This feature allows MEAM to model directional bonding [57]. Therefore, the overall formalism in MEAM closely resembles that of EAM (Eq. (1)), with a notable distinction lying in MEA's effective electron density, which includes angular contributions [57]:

$$\rho_i^{(0)} = \sum_j \phi^{(0)}(r_{ij}) \tag{3}$$

$$\left(\rho_i^{(1)}\right)^2 = \sum_{\alpha}\left[\sum_j \phi^{(1)}(r_{ij})\frac{\alpha_{ij}}{r_{ij}}\right]^2 \tag{4}$$

$$\left(\rho_i^{(2)}\right)^2 = \sum_{\alpha,\beta}\left[\sum_j \phi^{(2)}(r_{ij})\frac{\alpha_{ij}\beta_{ij}}{r_{ij}^2}\right]^2 - \frac{1}{3}\left[\sum_j \phi^{(2)}(r_{ij})\right]^2 \tag{5}$$

$$\left(\rho_i^{(3)}\right)^2 = \sum_{\alpha,\beta,\gamma}\left[\sum_j \phi^{(3)}(r_{ij})\frac{\alpha_{ij}\beta_{ij}\gamma_{ij}}{r_{ij}^3}\right]^2 - \frac{1}{3}\left[\sum_j \phi^{(2)}(r_{ij})\right]^2 \tag{6}$$

where $\phi^{(l)}$ represents the atomic electron densities weighted by the x, y and z components of the distances between atoms (identified by α, β, γ). Note that the spherically symmetric partial electron density $\rho_i^{(0)}$ is the same as the electron density in the EAM (Eq. (2)). Finally, it is worth mentioning that the MEAM, in contrast to EAM, is an empirical extension and has not been justified by strong physical arguments [58]. Though this potential has been used for different simulation systems, recent works usually avoid using it because it considers only first nearest-neighbor interactions using instead a modified version known as second nearest-neighbor modified embedded-atom method potential (2NN-MEAN) that rectifies this by incorporating second nearest-neighbor interactions [55].

2. Computational methodology

2.1 Processing

We conducted MD simulations using the open-source code LAMMPS. The atomic interactions in all the simulations were modeled using the MEAM interatomic potential, as parameterized by Huang et al. [59].

To accurately determine the lattice structure of the Hf-Nb-Ta-Ti-Zr equiatomic alloy, we created single-crystalline samples with BCC, FCC, and HCP structures.

Initially, these samples had distinct solid solution configurations composed by a total of 24,000 atoms, with 4800 atoms of each element. Periodic boundary conditions were applied in all simulations along the three orthogonal directions to eliminate surface effects. To obtain the equilibrated structure of the HEA, samples underwent energy minimization at 0 K using the conjugate gradient method. **Figure 1** depicts snapshots of the solid solution sample under a BCC structure captured in the stages before energy minimization (a), and after heating and stabilization at 300 K (b). The coloring follows the CNA [60] algorithm, enabling a clear observation of the structural differences of surface-defect atoms in gray color and volumetric ones in blue.

After the minimization stage to ensure the stable structure of HEA, the sample was submitted to heating processes until reaching room temperature, set as 300 K, with a heating rate of 1.0×10^{10} K/s. In this process, the isothermal-isobaric (NPT) ensemble, where temperature and pressure are controlled by the Nose-Hoover thermostat and barostat, was used. In a second step, another heating was carried out until the sample reached a temperature of 1500 K, while maintaining the same thermodynamic conditions. To achieve a stable structure at that temperature, a thermal stabilization of 0.1 ns was conducted under the control of the canonical (NVT) ensemble. After the heating and stabilization stages, the system underwent controlled cooling at a rate of 1×10^{10} K/s under the same conditions as during the heating, until reaching a temperature of 300 K. In addition, the sample was stabilized at this temperature for a period of 0.1 ns to remove possible residual stresses.

Once the equilibrium structure at room temperature was determined, single-crystal samples of the equiatomic Hf-Nb-Ta-Ti-Zr HEA, comprising a total of 113,800 atoms and arranged in BCC crystalline structure, were generated in a cylindrical shape with a diameter and height around of 10.1 and 27.2 nm. The effects of anisotropy on the mechanical behavior of the alloy were studied by adjusting the tensile efforts parallel to the z-axis along the [001], [110], and [111] crystallographic directions. **Figure 2** displays a snapshot of the samples in the initial stage of the simulation, before the energy minimization process.

Subsequently, the samples were thermalized and stabilized at 300 K for 0.1 ns under the control of the NPT ensemble. Samples underwent uniaxial tensile tests in the [001], [110], and [111] crystallographic directions at a strain rate of 1.0×10^9 s^{-1}

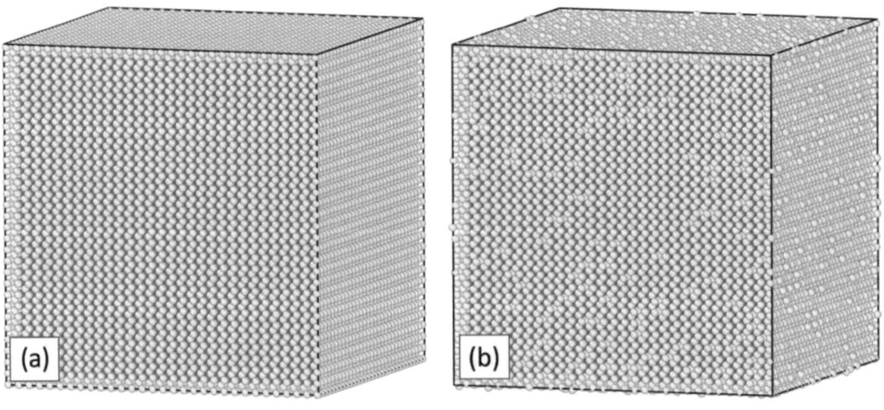

Figure 1.
Snapshot of BCC Hf-Nb-Ta-Ti-Zr HEA (a) before energy minimization step and, (b) after heating and stabilization at 300 K.

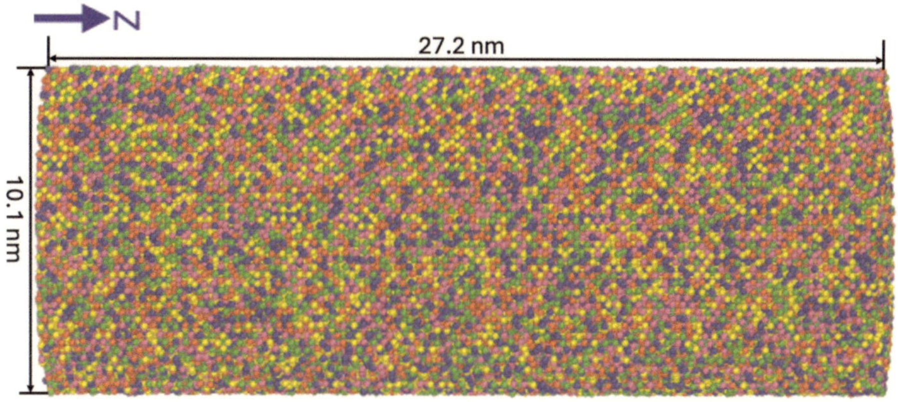

Figure 2.
Snapshots of cylindrical sample with the z-axis parallel to the displacement application. The atoms of Hf, Nb, Ta, Ti, and Zr are represented as spheres with colors red, blue, yellow, pink, and green, respectively.

and at 300 K. The stresses in the orthogonal directions to the tensile axis were set to zero during deformation.

2.2 Post-processing

The post-processing stage is extremely important in MD studies, producing results that correspond to the properties and characteristics of the materials. The structural characteristics of the material were analyzed using the algorithms: total pair distribution functions (PDF), common neighbor analysis (CNA) [60], and Warren-Cowlay (WC) [61]. In addition, the X-ray diffraction technique, widely used in experimental studies, was utilized in the simulations. The XRD parameters were set to consider radiation from copper Kα with a wavelength of 0.15418 nm and an angular 2θ range from 30 to 100 degrees. The post-processing analysis corresponding to mechanical deformation samples was carried out by applying the displacement extraction algorithm (DXA) [62], as implemented in the OVITO package [63].

3. Results and discussion

3.1 Analysis of the structure of Hf-Nb-Ta-Ti-Zr alloy

In the initial steps of DM simulation, the natural choice is to achieve thermodynamically and structurally stable systems. The structural stability strongly depends on the interatomic potential used, which is a key factor in producing reliable results. The stable structure of high entropy alloys is determined by performing energy minimization in the simulations, while also considering ideal lattices as BCC, FCC, and HCP. The simulations allow the determination of two factors of fundamental importance for the accuracy of the results: the cohesion energy, which depends on potential energy, and the lattice parameter a. As an example, **Figure 3(a)** shows the curves of potential energy (PE) as a function of the atomic separation distance (r) for the BCC crystalline sample. **Figure 3(b)** depicts curves fitted to different sets of data obtained from energy minimization stages. It should be noted that the BCC structure produced the

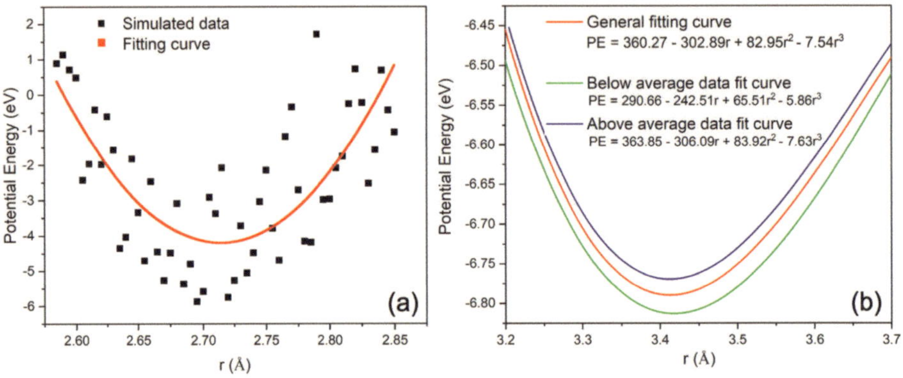

Figure 3.
(a) Potential energy (PE) curve as a function of atomic separation distance (r) for HEA sample with BCC crystalline lattice. (b) Fitted curves to different sets of data obtained from energy minimization stages.

lowest PE value among the other structures, then this structure was considered as the stable equilibrium of the equiatomic Hf-Nb-Ta-Ti-Zr HEA.

During the energy minimization stage, the PE vs. r curves generally show data with minimal dispersion in the energy, with increasing or decreasing in the separation distances. However, in the present case, a large dispersion in the PE values was found, which is unexpected, despite the alloy having several components that typically lead to strong distortion of the crystalline lattice. A non-linear fit of the data using a third-order polynomial equation, typically used in similar cases, yields a curve described by the eq. $PE = 360.27–302.8r + 82.95r^2–7.54r^3$. The fitting presents a correlation factor (R^2) of 0.93, with energy and lattice parameter of −7.010 eV and 3.425 Å, respectively. The estimated standard deviation was around 3%. Furthermore, it can be observed that the fitting curve exhibits two distinct envelopes of data: one above the average fitting curve and the other below it. Taking into account the observable scattering, we considered a possible mixture of phases with BCC lattices. Subsequently, the data were separated into two groups, and new fittings were made, under the same conditions as those for the initial fitting. It is worth noting that this situation is not typical, and great care is required for its use. **Figure 3(b)** shows both curves, labeled as BCC-1 for the data located above the average fitting curve and BCC-2 for the data located below that curve. The fittings of the two set of data result in the relationship: $PE = 363.85–306.0r + 83.92r^2–7.63r^3$ for BCC-1 and $PE = 290.66–242.5r + 65.51r^2–5.86r^3$ for BCC-2. The potential energy and lattice parameter values are shown in **Table 1**. Both fittings exhibit a correlation factor of 0.98, with an estimated standard deviation of less than 2%.

The results of **Table 1** align well with those obtained by Huang et al. [60], who reported a BCC structure for the equiatomic alloy in the same system. However, the simulated values deviate from those obtained by experimental XRD of samples produced by powder metallurgy [64] and subjected to Rietveld refinement. The

Lattice structure	BCC	BCC-1	BCC-2
Potential energy (eV)	−7.010	−6.769	−6.813
Lattice parameter (Å)	3.425	3.413	3.418

Table 1.
Potential energy and lattice parameters for BCC lattices in thermodynamic equilibrium state.

measured lattice parameter of real samples is 3.4024 Å for the same equiatomic HEA. This slight difference is expected due to variations in methods, such as the system size [65–67] and the inherent variability in the typical characterization of real samples. Thus, the values obtained in this work by MD are considered acceptable when compared with the experimental data. Furthermore, the good correlation in the fittings of the separated curves indicates a high probability of the material presenting BCC phase separation. This hypothesis was corroborated by analyzing of the simulated XRD and comparing it to experimental data reported by Málek et al. [64].

Figure 4 shows the X-rays diffractograms (XRD) obtained from experiments and simulations under two temperatures. The experimental data were obtained by capturing the XRD curves from Málek et al. [64] using a custom Python code. At 300 K, the simulated diffractogram shows diffraction peaks corresponding to both BCC phases, as determined by Lukac et al. [68], with lattice parameters of a = 3.3197 Å for the BCC-1 phase, and a = 3.4181 Å for the BCC-2. However, the simulated curves do not show the corresponding peaks for the experimentally determined HCP phase.

In **Figure 5(a)**, the curve of PE vs. temperature is depicted during the heating stage of the equiatomic Hf-Nb-Ta-Ti-Zr HEA. Two approximately linear regions emerge, each characterized by distinct slopes. The inflection point occurs at a temperature around 1100 K, suggesting that a solid-state phase transformation takes place. As the alloy was only heated, it is supposed that the phase transition is likely thermally activated. However, it occurs in a short range of temperatures and very quickly, which allows us to infer that this phase transition is similar to a martensitic transformation.

A clearer observation can be made in the curves of **Figure 5(b)**, which display the evolution of BCC, FCC, and HCP structures, along with other atomic configurations following the CNA algorithm. The CNA algorithm can detect various atomic structures, including BCC, FCC, HCP, icosahedral, and other structural configurations. These additional configurations mainly correspond to certain crystalline defects, such as dislocations, twins, and surface atoms, as illustrated in **Figure 1**, where blue-colored atoms represent volumetric atoms and gray atoms represent surface atomic configurations, among other defects. In **Figure 5(b)**, as the temperature increases, it is possible to observe a rapid decrease in the volume of BCC structures, accompanied by a fast increase in other types of structures. Additionally, at temperatures around

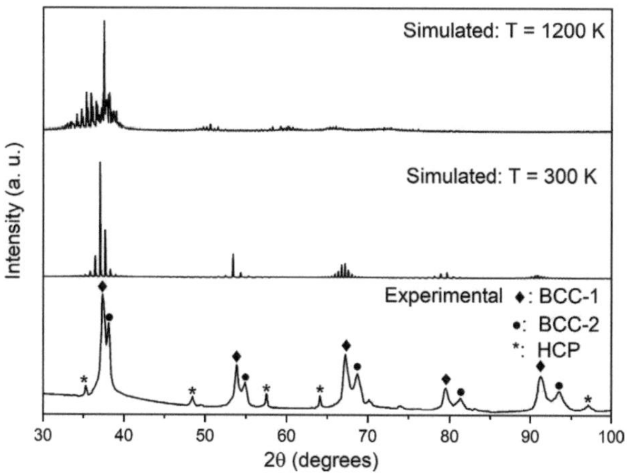

Figure 4.
X-ray diffractograms at temperatures of 300, and 1200 K together with the curves obtained experimentally by Málek et al. [366] for the Hf-Nb-Ta-Ti-Zr HEA.

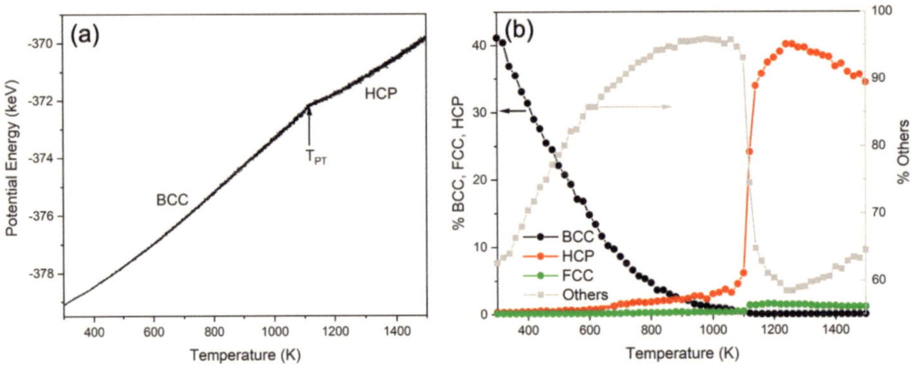

Figure 5.
(a) Potential energy vs. temperature during the heating stage at a rate of 1×10^{10} K/s of the Hf-Nb-Ta-Ti-Zr alloy. (b) Evolution of BCC, FCC, HCP, and other atomic configuration according to CNA algorithm. Both set of data show the fast phase transition from BCC to HCP structures. T_{PT} corresponds to the temperature of phase transition.

1100 K, the population of the BCC structure decreases to zero, while the HCP structure increases its population to about 37%.

On the other hand, the application of the DXA algorithm at the evolution of the atomic structure in function of temperature shows that initially at 300 K, there are a small population of ½<100 > dislocations. When the sample is stressed, the dislocation density increases rapidly, however, the atom population involved in the other type of structures is higher than that of the involved in the dislocations. Thus, we can infer that many atom fractions are involved in defects such as stacking faults and twins. In temperatures close to the T_{PT} the density of ½<100 > dislocations becomes to zero, nevertheless, a small fraction of the HCP phase starts to form and its fraction increase rapidly giving rise to the formation of $1/3<1\bar{1}00>$ dislocation.

Figure 6 depicts the analysis of small range order (SRO) by using the Warren-Cowley (WC) parameters (α_{ij}) in the quinary Hf-Nb-Ta-Ti-Zr HEA after the stabilization stage at 300 K. The α_{ij} parameters are determined around an atom *i* within the first nearest-neighbor shell [59]. A positive value indicates that the atom pair is unfavored while for a negative value the atom pair is preferred. The formation of Hf-Nb, Nb-Ta, and Hf- Zr atom pairs are favored. Other atom pairs are not observed, which means that these pairs are randomly distributed inside the material. It is worth noting that in spite of the little difference in fraction of Hf-Nb pairs this result is in

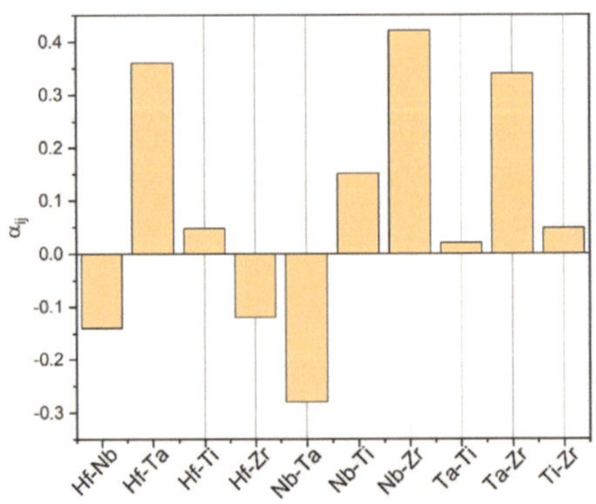

Figure 6.
Warren-Cowley parameters for determining the atom pairs formation.

agreement with that obtained by Huang et al. [59] using a hybrid Monte Carlo MC/ MD simulation. The authors found that in the quinary HEA the main atom pairs are Hf-Nb, Nb-Ta, and Hf-Zr.

The population of surface atoms on an ideal defect-free sample corresponds to 18.4%. However, when stabilized at 300 K this fraction increases to 22.4%. Consequently, the actual change in the atom population is approximately 15%. On the other hand, the DXA analysis shows that the increase in the population of other atomic configurations is not associated with dislocation formation. In fact, at 300 K, the sample presents just a few <100> dislocations, and this number decreases to zero as the temperature increase. Thus, the increase in atom population associated with other types of structures may be attributed to the formation of twins, which are not clearly identified by the DXA algorithm.

Furthermore, as the material undergoes a thermally induced transformation, it is crucial to determine the crystalline structure at high temperatures. To identify that structure, the sample obtained after the phase transition was subjected to an XRD test, specifically at a temperature of 1200 K. The diffractogram indicates a mixture of crystalline phases, with the matrix corresponding to an HCP structure, while a fraction of the remaining material shows a BCC-1 as the secondary phase (see **Figure 4**).

In general, metallic materials tend to expand when subjected to heat. Thermal expansion is usually observed as a straight line with a certain slope that indicates the volumetric expansion modulus. In the case of PE vs. T curves, at temperatures below as well as above 1100 K, there is a slight curvature instead of the expected linearity. The curvatures around 1100 K occur due to the mixing of two different phases. At low temperatures, starting from room temperature, the material is composed by a mixture of the two BCC phases (BCC-1 and BCC-2) with very close lattice parameters. Above 1100 K, the phases present are HCP and BCC-2. Although the alloy processed by Lukac et al. [68] using spray technique has both cubic phases, when the powder is consolidated and sintered at temperatures above 1400 K, the alloy no longer exhibits the HCP phase found by MD.

The structural analysis by using PDFs enables us to infer interesting characteristics of the alloy. **Figure 7** shows the PDF curves at 0 K, obtained after energy minimization, 300 and 1200 K. The PDF curves show peaks corresponding to the first, second, third, etc., nearest neighbors when considering a central atom as a reference. The curves have the well-established shape characteristic of the BCC structure. Moreover, there are small peaks that are not typically observed in the PDF patterns of BCC structures in pure

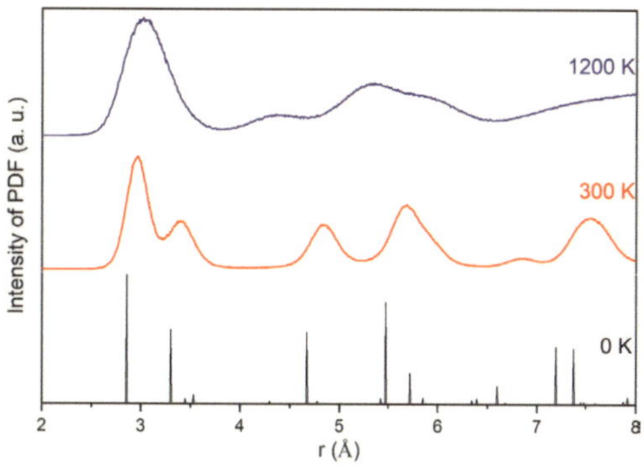

Figure 7.
Pair distribution functions of the Hf-Nb-Ta-Ti-Zr HEA, at temperatures of 0, 300, and 1200 K.

metals. These peaks may be related to the surface atoms, or possibly linked to other structures present in the material at a small volume fraction. At the temperature of 300 K, the PDF curves show low-intensity peaks with an elongated base, indicating a change in atomic separation distances from the initial structure at 0 K. This change is associated with crystalline lattice distortion and possibly the formation of the second cubic phase. Indeed, the formation of approximately 0.2% of the HCP phase which was detected by the CNA algorithm. At 1200 K, there is a decrease in the intensity of the peaks and a wider base, particularly noticeable in the first peak where there is a coalescence of the two peaks. In addition, the secondary peaks shift to other atomic separation distances, suggesting the formation of a second phase.

The phase transformation from BCC to HCP structures was observed during heating, considering that atomic diffusion in HEAs is relatively slow. Hence, it is thought that the nucleation of the phase transformation caused lattice distortions that influenced the elastic deformation. It should be noted that there is a high percentage of unidentified structures at both temperatures of 300 and 1200 K, constituting 22.4 and 37.0% of the total atoms, respectively. At 0 K, the fraction was of 18.4%, primarily located on the surfaces of the sample, possibly associated with surface defects as these atoms are not fully bonded. Considering that the 18.4% fraction is attributable to the imperfections in bonding, the atomic fraction immersed in the unidentified structures was of 4.0 and 18.6% at 300 and 1200 K.

3.2 Analysis of the mechanical behavior of Hf-Nb-Ta-Ti-Zr alloy

The technological applications of any material strongly depend on its physical, chemical and mechanical properties. For instance, in structural materials, mechanical properties are crucial. Therefore, it is very important to determine their responses to various forms of external stress or forces, such as tensile, compression, shear, or flexural loads. To determine the mechanical properties in metallic materials, it is customary to characterize by tensile tests, then this technique was adopted here.

In the present work, we investigated the effect of single-crystalline directions on the mechanical properties of the Hf-Nb-Ta-Ti-Zr HEA. The elongation was applied in three crystallographic directions, namely [001], [110], and [111] of a BCC structure. **Figure 8** displays the stress (σ)–strain (ε) curves obtained from MD simulations in the

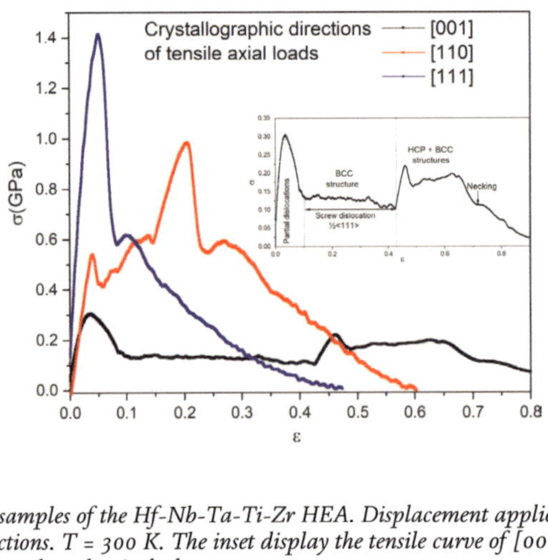

Figure 8.
σ–ε curves in single-crystal samples of the Hf-Nb-Ta-Ti-Zr HEA. Displacement applied in the [001], [110], and [111] crystallographic directions. T = 300 K. The inset display the tensile curve of [001] load direction with different stages of structures and mechanical phenomena.

conditions where the anisotropic effects are clearly established. Thus, the values of ultimate stress (σ_u), elastic modulus (E), and elongation of the same HEA alloy vary with the direction of elongation. The inset of **Figure 8** shows the structural evolution of sample in [0 0 1] direction.

For the BCC Hf-Nb-Ta-Ti-Zr HEA, the [111] direction represents higher compactness, hence the resistance of the sample to deformation exhibits higher resistance. The lowest mechanical strength is observed when the deformation is applied in the [001] direction, reaching an ultimate stress of approximately 300 MPa. Subsequently, the material undergoes a softening, with the stress stabilizing at around 130 MPa. However, the material exhibits a pseudo-elastic behavior with ductility close to 43%. A high strain induces a phase transformation in which the material hardens slightly, reaching an σ_u of 220 MPa, followed by pseudo-elastic behavior up to approximately 70% strain.

Figure 9 depicts the evolution in atomic configurations during the tensile test with a load applied in [001] crystallographic direction. **Figure 9(a)** shows a snapshot of the unloaded cylindrical sample, i.e., before the tensile test. **Figure 9(b)** depicts the sample at the instant when the σ_u is achieved. According to the DXA algorithm, at that point, there are some 1/6 ⟨112⟩ partial Shockley dislocations [69], which can give rise to the formation of deformation twins. As the deformation increases and reaches the σ_u, there is a little increase in the density of partial Shockley dislocations, and this phenomenon is accompanied by the appearance of some Hirth 1/3 ⟨100⟩ [70], and a great number of the screw ½<111 > dislocations. In single-component BCC systems, the shortest Burgers vector correspond to the perfect ½<111 > dislocation. In the present work, screw dislocations are present in the sample after the initial stage of homogeneous deformation. With the increase in deformation, the partial dislocations are no longer detected, also, the screw dislocation population decreases. However, according to the DXA algorithm the structure is predominantly BCC until 43% of strain. At that point, the HCP phase starts to appear and the structure increases rapidly, however, there is a remanent fraction of the BCC structure. Furthermore, the sample undergoes a change in shape from circular to elliptical, as shown in **Figures 9(c)** and **(d)**. The elliptical shape persists during homogeneous pseudo-elastic

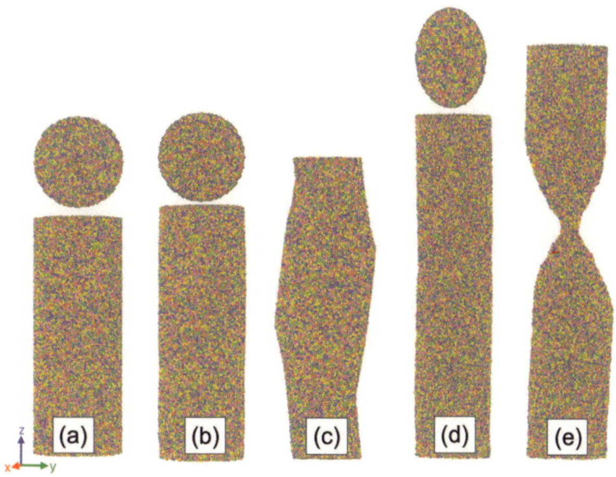

Figure 9.
Shapes of Hf-Nb-Ta-Ti-Zr sample during uniaxial tensile test along the [001] direction: (a) before the test, (b) at ultimate stress, (c) onset of elliptical shape, (d) full elliptical shape, and (e) close to fracture.

deformation and is sustained during the fracture process of the material, as depicted in **Figure 9**. Also, a significant level of deformation leads to the formation of a very thin neck, indicative of high ductility.

The analysis of the atomic structure under deformation by using the CNA tool indicates a continuous variation in the atomic fraction belonging to the BCC or HCP structures. Initially, the entire sample is composed mainly of a BCC matrix phase, and with deformation progress, part of it is continuously transformed into the HCP structure. This result is corroborated by the DXA results that show that during the first stage of strain the structure is predominantly BCC, but the phase transformation to HCP structure occurs at 43% of strain, as can be seen on the inset of **Figure 8**. During a short range of strain, the HCP phase reaches a high fraction, even though the BCC phase still remains as a minority phase. Its evolution persists until the formation of the neck. This phase transformation significantly influences the mechanical behavior of the HEA.

It should be noted that the main dislocation type formed during the deformation process of the BCC phases is the ½ < 111>. After the phase transformation, two distinct types of HCP dislocation emerge: $1/3<\mathbf{1\bar{2}10}>$ and $1/3<\mathbf{1\bar{1}00}>$, which are not observed in the case of the thermally induced phase transformation.

When the deformation is applied in the [110] direction, there is an increase in the σ_u, reaching close to 540 MPa. However, after a slight drop in the stress, there is continuous hardening until it reaches 980 MPa. This fact is related to the formation of a large number of dislocations, leading to a new phase induced by deformation. The material then undergoes continuous softening until fracture, initiating at approximately 30% of deformation.

The application of the tensile efforts in the [111] crystallographic direction produces a higher ultimate stress, exceeding 1400 MPa. This stress is the highest value reached among the three studied directions. Nevertheless, this elevated stress level is associated with a significant reduction in ductility. In addition, the stress drop occurs relatively fast, when the phase transformation is not well defined, with the stress-strain halo corresponding to the second phase being almost imperceptible. In this configuration, there was no observed presence of necking, and the sample failed with a low ductility.

The ultimate stress, relative ductility, and elastic modulus values corresponding to each crystallographic direction are summarized in **Table 2**. The anisotropic effect is clearly detected by the MD simulation for the monocrystal. Nevertheless, the elastic modulus is very low, with values lower than those of pure aluminum (E = 61 GPa) and magnesium (E = 44 GPa) [71]. However, it should be noted that these moduli correspond to polycrystalline materials, while in the current work, values are reported for three crystallographic directions of a single crystalline sample. Moreover, the simulated results are always influenced by the strong impact of the chosen

Direction	**[001]**		**[110]**		**[111]**
Phase	**BCC**	**HCP**	**BCC**	**HCP**	**BCC**
σ_u (MPa)	303	219	538	981	1420
Elongation (%)	42	28	14	21	15
E (GPa)	11.0	3.4	16.4	7.8	28.9

Table 2.
Mechanical properties obtained from the tensile curves at 300 K.

interatomic potential. Therefore, any results must be interpreted with a certain level of confidence.

The tensile test applied to different crystallographic directions at room temperature shows distinct mechanical behavior. The BCC to HCP phase transformation induced by strain is strongly dependent on the crystal orientation. The [001] uniaxial tensile test produces a lower density of dislocation leading to a significant BCC to HCP transformation, probably due to the emissions of dislocations and twins. Uniaxial test applied in [110] direction allows a partial phase transformation; however, a density of dislocation increases in respect to the [001] direction. The tensile test in the [111] direction reveals a higher dislocation density, with no noted phase transformation. Those results are in good agreement with the study of Hsieh et al. [72], who reported phase transformation induced by stress on Co-Cr-Fe-Mn-Ni HEA with crystal orientation dependence.

4. Conclusions

The equiatomic Hf-Nb-Ta-Ti-Zr high entropy alloy was studied by classic molecular dynamics simulation in a single crystal. The structural features and the mechanical properties were simulated using the LAMMPS code and partial data analysis was carried out in Ovito code. At 300 K, the structure is composed by two BCC structures with very similar lattice parameters. Heating from 300 to 1200 K induces a phase transformation from BCC to HCP structures. The tensile mechanical test used a round sample with a diameter of 10.1 nm and a length of 27.2 nm. The samples were elongated at a strain rate of $1.0 \times 10^9\ s^{-1}$ in three crystallographic directions: [001], [110], and [111]. The results revealed a strong anisotropy in mechanical properties, with the [111] direction exhibiting greater resistance. The tensile test in [111] direction showed a higher dislocation density; however, no phase transformation was observed during the tensile test in this direction. Moreover, the phase transformation is strongly dependent on the crystallographic direction.

Acknowledgements

This study was financed in part by the Coordenação de Aperfeiçoamento de Pessoal de Nível Superior – Brasil (CAPES) – Finance Code 001. The authors also are grateful to FAPERJ – Fundação Carlos Chagas Filho de Amparo à Pesquisa do Estado do Rio de Janeiro (Grants: SEI-260003/014215/2021 and SEI-260003/001582/2022) and CNPq (Grant: 303023/2019-8).

Author details

Luis César R. Aliaga*, Alexandre Melhorance Barboza, Loena Marins de Couto and Ivan Napoleão Bastos
Polytechnic Institute, Rio de Janeiro State University, Nova Friburgo, Rio de Janeiro, Brazil

*Address all correspondence to: aliaga@iprj.uerj.br

References

[1] Babić E, Drobac D, Figueroa IA, Laurent-Brocq M, Marohnić Ž, Trontl VM, et al. Transition from high-entropy to conventional alloys: Which are better? Materials. 2021;**14**(19):5824. DOI: 10.3390/ma14195824

[2] Zhang W, Liaw PK, Zhang Y. Science and technology in high-entropy alloys. Science China Materials. 2018;**61**(1): 2-22. DOI: 10.1007/s40843-017-9195-8

[3] Yeh JW, Chen SK, Lin SJ, Gan JY, Chin TS, Shun TT, et al. Nanostructured high-entropy alloys with multiple principal elements: Novel alloy design concepts and outcomes. Advanced Engineering Materials. 2004;**6**:299-303. DOI: 10.1002/adem.200300567

[4] Cantor B, Chang ITH, Knight P, Vincent AJN. Microstructural development in equiatomic multicomponent alloys. Materials Science and Engineering A. 2004;**375–377**:213-218. DOI: 10.1016/j.msea.2003.10.257

[5] Jin K, Bei H. Single-phase concentrated solid-solution alloys: Bridging intrinsic transport properties and irradiation resistance. Frontiers in Materials. 2018;**5**:26. DOI: 10.3389/fmats.2018.00026

[6] Miracle DB. High entropy alloys as a bold step forward in alloy development. Nature Communications. 2019;**10**:1805. DOI: 10.1038/s41467-019-09700-1

[7] Sharma P, Dwivedi VK, Dwivedi SP. Development of high entropy alloys: A review. Materials Today: Proceedings. 2021;**43**:502-509. DOI: 10.1016/j.matpr.2020.12.023

[8] George EP, Raabe D, Ritchie RO. High-entropy alloys. Nature Reviews Materials. 2019;**4**:515-534. DOI: 10.1038/s41578-019-0121-4

[9] George EP, Curtin WA, Tasan CC. High entropy alloys: A focused review of mechanical properties and deformation mechanisms. Acta Materialia. 2020;**188**: 435-474. DOI: 10.1016/j.actamat.2019.12.015

[10] Yang T, Zhao Y, Liu W, Kai J, Liu C. L12-strengthened high-entropy alloys for advanced structural applications. Journal of Materials Research. 2018;**33**: 2983-2997. DOI: 10.1557/jmr.2018.186

[11] Joele M, Matizamhuka WR. A review on the high temperature strengthening mechanisms of high entropy superalloys (HESA). Materials. 2021;**14**(19):5835. DOI: 10.3390/ma14195835

[12] Yin Y, Ren W, Tan Q, Chen H, Huang H, Zhang MX. A cost-effective cryogenic high-entropy alloy with high strength-ductility synergy and strain hardenability. Materials Science and Engineering A. 2023;**865**:144607. DOI: 10.1016/j.msea.2023.144607

[13] Marques F, Balcerzak M, Winkelmann F, Zepon G, Felderhof M. Review and outlook on high-entropy alloys for hydrogen storage. Energy & Environmental Science. 2021;**14**: 5191-5227. DOI: 10.1039/d1ee01543e

[14] Feng J, Tang Y, Liu J, Zhang P, Liu C, Wang L. Bio-high entropy alloys: Progress, challenges, and opportunities. Frontiers in Bioengineering and Biotechnology. 2022;**10**:977282. DOI: 10.3389/fbioe.2022.977282

[15] Yeh JW, Lin SJ. Breakthrough applications of high-entropy materials. Journal of Materials Research. 2018;**33**: 3129-3137. DOI: 10.1557/jmr.2018.283

[16] Ren JT, Chen L, Wang HY, Yuan ZY. High-entropy alloys in electrocatalysis: From fundamentals to applications. Chemical Society Reviews. 2023;**52**: 8319-8373. DOI: 10.1039/D3CS00557G

[17] Chang X, Zeng M, Liu K, Fu L. Phase engineering of high-entropy alloys. Advanced Materials. 2020;**32**(14): 1907226. DOI: 10.1002/adma. 201907226

[18] Zhang Y. High-Entropy Materials – A Brief Introduction. Singapore: Springer; 2019. 152 p. DOI: 10.1007/ 978-981-13-8526-1

[19] Schneider M, Laplanche G. Effects of temperature on mechanical properties and deformation mechanisms of the equiatomic CrFeNi medium-entropy alloy. Acta Materialia. 2021;**204**:116470. DOI: 10.1016/j.actamat.2020.11.012

[20] Kozak R, Sologubenko A, Steurer W. Single-phase high-entropy alloys – An overview. Zeitschrift für Kristallographie - Crystalline Materials. 2015;**230**(1):55-68. DOI: 10.1515/zkri-2014-1739

[21] Pradeep KG, Tasan CC, Yao MJ, Deng Y, Springer H, Raabe D. Non-equiatomic high entropy alloys: Approach towards rapid alloy screening and property-oriented design. Materials Science and Engineering A. 2015;**648**: 183-192. DOI: 10.1016/j.msea.2015. 09.010

[22] Su Z, Quan Z, Shen T, Jin P, Li J, Hu S, et al. A novel BCC-structure Zr-Nb-Ti medium-entropy alloys (MEAs) with excellent structure and irradiation resistance. Materials. 2022;**15**:6565. DOI: 10.3390/ma15196565

[23] Hatzenbichler L, Zeisl S, Clemens H, Holec D. Phase stability of TiAl-based BCC high entropy alloys. Intermetallics. 2023;**158**:107893. DOI: 10.1016/j.inter met.2023.107893

[24] Onawale OT, Cobbinah PV, Nzeukou RA, Matizamhuka WR. Synthesis route, microstructural evolution, and mechanical property relationship of high-entropy alloys (HEAs): A review. Materials. 2021; **14**(11):3065. DOI: 10.3390/ ma14113065

[25] Li W, Liu P, Liaw PK. Microstructures and properties of high-entropy alloy films and coatings: A review. Materials Research Letters. 2018; **6**(4):199-229. DOI: 10.1080/21663831. 2018.1434248

[26] Gao X, Chen R, Liu T, Fang H, Qin G, Su Y, et al. High-entropy alloys: A review of mechanical properties and deformation mechanisms at cryogenic temperatures. Journal of Materials Science. 2022;**57**:6573-6606. DOI: 10.1007/s10853-022-07066-2

[27] Laurent-Brocq M, Mereib D, Garcin G, Monnier J, Perrière L, Villeroy B. Chemical architecturation of high entropy alloys through powder metallurgy. Journal of Alloys and Compounds. 2020;**835**:155279. DOI: 10.1016/j.jallcom.2020.155279

[28] Jiang D, Xie L, Wang L. Current application status of multi-scale simulation and machine learning in research on high-entropy alloys. Journal of Materials Research and Technology. 2023;**26**:1341-1374. DOI: 10.1016/ j.jmrt.2023.07.233

[29] Toda-Caraballo I, Wróbel JS, Nguyen-Manh D, Pérez P, Rivera-Díaz-del-Castillo PEJ. Simulation and modeling in high entropy alloys. The Journal of the Minerals, Metals & Materials Society. 2017;**69**:2137-2149. DOI: 10.1007/s11837-017-2524-2

[30] Xu S, Jian WR, Beyerlein IJ. Ideal simple shear strengths of two HfNbTaTi-based quinary refractory multi-principal element alloys. APL Materials. 2022;**10**: 111107. DOI: 10.1063/5.0116898

[31] Mills LH, Emigh MG, Frey CH, Philips NR, Murray SP, Shin J, et al. Temperature-dependent tensile behavior of the HfNbTaTiZr multi-principal element alloy. Acta Materialia. 2023;**245**: 118618. DOI: 10.1016/j.actamat.2022.118618

[32] Zlotea C, Sow MA, Ek G, Couzinie JP, Perriere L, Guillot I, et al. Hydrogen sorption in TiZrNbHfTa high entropy alloy. Journal of Alloys and Compounds. 2019;**775**:667-674. DOI: 10.1016/j.jallcom.2018.10.108

[33] Ohtani H. The CALPHAD method. In: Czichos H, Saito T, Smith L, editors. Springer Handbook of Materials Measurement Methods. Berlin: Springer; 2007. pp. 1001-1030. DOI: 10.1007/978-3-540-30300-8

[34] Tian F. A review of solid-solution models of high-entropy alloys based on ab-initio calculations. Frontiers in Materials. 2017;**4**:36. DOI: 10.3389/fmats.2017.00036

[35] Liu X, Zhang J, Yin J, Bi S, Eisenbach M, Wang Y. Monte Carlo simulation of order-disorder transition in refractory high entropy alloys: A data-driven approach. Computational Materials Science. 2021;**187**:110135. DOI: 10.1016/j.commatsci.2020.110135

[36] Kushnerov OI. Molecular dynamics simulation of the solidification of AlCoCuFeNi high–entropy alloy nanowire. Journal of Physics and Electronics. 2019;**27**(2):61-64. DOI: 10.15421/331925

[37] Ye L, Ming L, Hongyu L. Molecular dynamics study of local structure evolution of AlCrCoCuFeNi high entropy alloy in rapidly cooled process. Journal of Computational and Theoretical Nanoscience. 2015;**12**(11): 4649-4653. DOI: 10.1166/jctn.2015.4413

[38] Fernández-Caballero A, Wróbel JS, Mummery PM, Nguyen-Manh D. Short-range order in high entropy alloys: Theoretical formulation and application to Mo-Nb-Ta-V-W system. Journal of Phase Equilibria and Diffusion. 2017;**38**:391-403. DOI: 10.1007/s11669-017-0582-3

[39] Kostiuchenko T, Körmann F, Neugebauer J, Shapeev A. Impact of lattice relaxations on phase transitions in a high-entropy alloy studied by machine-learning potentials. npj Computational Materials. 2019;**5**:55. DOI: 10.1038/s41524-019-0195-y

[40] Liu X, Zhang J, Pei Z. Machine learning for high-entropy alloys: Progress, challenges and opportunities. Progress in Materials Science. 2023;**131**: 101018. DOI: 10.1016/j.pmatsci.2022.101018

[41] Rao Z, Tung PY, Xie R, Wei Y, Zhang H, Ferrari A, et al. Machine learning–enabled high-entropy alloy discovery. Science. 2022;**378**(6615): 78-85. DOI: 10.1126/science.abo4940

[42] Roy D, Mandal SC, Pathak B. Machine learning assisted exploration of high entropy alloy-based catalysts for selective CO2 reduction to methanol. The Journal of Physical Chemistry Letters. 2022;**13**(25):5991-6002. DOI: 10.1021/acs.jpclett.2c00929

[43] Gorsse S, Nguyen MH, Senkov ON, Miracle DB. Database on the mechanical properties of high entropy alloys and complex concentrated alloys. Data in Brief. 2018;**21**:2664-2678. DOI: 10.1016/j.dib.2018.11.111

[44] Thompson AP, Aktulga HM, Berger R, Bolintineanu DS, Brown WM, Crozier PS, et al. LAMMPS – A flexible simulation tool for particle-based materials modeling at the atomic, meso, and continuum scales. Computer Physics Communications. 2022;**271**:108171. DOI: 10.1016/j.cpc.2021.108171

[45] Plimpton S. Fast parallel algorithms for short-range molecular dynamics. Journal of Computational Physics. 1995; **117**(1):1-19. DOI: 10.1006/jcph.1995.1039

[46] Müser MH, Sukhomlinov SV, Pastewka L. Interatomic potentials: Achievements and challenges. Advances in Physics X. 2022;**7**(1):2093129. DOI: 10.1080/23746149.2022.2093129

[47] Daw MS, Baskes MI. Semiempirical, quantum mechanical calculation of hydrogen embrittlement in metals. Physical Review Letters. 1983;**50**:1285. DOI: 10.1103/PhysRevLett.50.1285

[48] Daw MS, Baskes MI. Semiempirical embedded-atom method: Derivation and application to impurities, surfaces, and other defects in metals. Physical Review B. 1984;**29**:6443. DOI: 10.1103/PhysRevB.29.6443

[49] Schommers W. Basic Physics of Nanoscience – Traditional Approaches and New Aspects at the Ultimate Level. Netherlands: Elsevier; 2019. 290 p. DOI: 10.1016/C2016-0-03658-6

[50] Vella JR, Stillinger FH, Panagiotopoulos AZ, Debenedetti PG. A comparison of the predictive capabilities of the embedded-atom method and modified embedded-atom method potentials for lithium. The Journal of Physical Chemistry B. 2015;**119**(29): 8960-8968. DOI: 10.1021/jp5077752

[51] Tschopp MA, Spearot DE, McDowell DL. Influence of grain boundary structure on dislocation nucleation in FCC metals. In: Hirth JP, Kubin L, editors. Dislocations in Solids. Netherlands: Elsevier; 2008. 310 p. DOI: 10.1016/S1572-4859(09)01509-5

[52] Kong J, Luo G, Tian Y, Du C. Atomic insight into tribological behavior of AlCoCrFeNi high entropy alloy at various nanoscratching conditions. Journal of Materials Research and Technology. 2023;**27**:7293-7303. DOI: 10.1016/j.jmrt.2023.11.150

[53] Yang X, Zhang J, Sagar S, Dube T, Kim BG, Jung YG, et al. Molecular dynamics modeling of mechanical and tribological properties of additively manufactured AlCoCrFe high entropy alloy coating on aluminum substrate. Materials Chemistry and Physics. 2021; **263**:124341

[54] Xie L, Brault P, Thomann AL, Bauchire JM. AlCoCrCuFeNi high entropy alloy cluster growth and annealing on silicon: A classical molecular dynamics simulation study. Applied Surface Science. 2013;**285**: 810-816. DOI: 10.1016/j.apsusc.2013.08.133

[55] Lee BJ, Baskes MI. Second nearest-neighbor modified embedded-atom-method potential. Physical Review B. 2000;**62**(13):8564-8567. DOI: 10.1103/PhysRevB.62.8564

[56] Baskes MI. Application of the embedded-atom method to covalent materials: A semiempirical potential for silicon. Physical Review Letters. 1987;**59**: 2666. DOI: 10.1103/PhysRevLett.59.2666

[57] Ackland GJ, Bonny G. Interatomic potential development. In: Konings RJM, Stoller RE, editors. Comprehensive Nuclear Materials. 2nd ed. Netherlands: Elsevier; 2019. pp. 544-572.

DOI: 10.1016/B978-0-12-803581-8.11687-X

[58] Baskes MI. Modified embedded-atom potentials for cubic materials and impurities. Physical Review B. 1992; **46**(5):2727-2742. DOI: 10.1103/PhysRevB.46.2727

[59] Huang X, Liu L, Duan X, Liao W, Huang J, Sun H, et al. Atomistic simulation of chemical short-range order in HfNbTaZr high entropy alloy based on a newly-developed interatomic potential. Materials & Design. 2021;**202**: 109560. DOI: 10.1016/j.matdes.2021.109560

[60] Faken D, Jónsson H. Systematic analysis of local atomic structure combined with 3D computer graphics. Computational Materials Science. 1994; **2**(2):279-286. DOI: 10.1016/0927-0256(94)90109-0

[61] Rao Y, Curtin WA. Analytical models of short-range order in FCC and BCC alloys. Acta Materialia. 2022;**226**: 117621. DOI: 10.1016/j.actamat.2022.117621

[62] Stukowski A, Bulatov VV, Arsenlis A. Automated identification and indexing of dislocations in crystal interfaces. Modelling and Simulation in Materials Science and Engineering. 2012; **20**(8):085007. DOI: 10.1088/0965-0393/20/8/085007

[63] Stukowski A. Visualization and analysis of atomistic simulation data with OVITO - The open visualization tool. Modelling and Simulation in Materials Science and Engineering. 2009;**18**(1):015012. DOI: 10.1088/0965-0393/18/1/015012

[64] Málek J, Zýka J, Lukáč F, Vilémová M, Vlasák T, Čížek J, et al. The effect of processing route on properties of HfNbTaTiZr high entropy alloy. Materials. 2019;**12**:4022. DOI: 10.3390/ma12234022

[65] Mahata A, Zaeema MA. Size effect in molecular dynamics simulation of nucleation process during solidification of pure metals: Investigating modified embedded atom method interatomic potentials. Modelling and Simulation in Materials Science and Engineering. 2019; **27**:085015. DOI: 10.1088/1361-651X/ab4b36

[66] Barboza AM, Bastos IN, Aliaga LCR. Molecular dynamics simulations of the mechanical behavior of nanostructured and amorphous Al80Ti15Ni5 alloy. Revista Facultad de Ingeniería Universidad de Antioquia. 2022;**103**: 20-33. DOI: 10.17533/udea.redin.20201009

[67] Aliaga LCR, Lima LVPC, Domingues GMB, Bastos IN, Evangelakis GA. Experimental and molecular dynamics simulation study on the glass formation of Cu–Zr–Al alloys. Materials Research Express. 2019;**6**: 045202. DOI: 10.1088/2053-1591/aaf97e

[68] Lukac F, Dudr M, Musalek R, Klecka J, Cinert J, Cizek J, et al. Spark plasma sintering of gas atomized high-entropy alloy HfNbTaTiZr. Journal of Materials Research. 2018;**33**:3247-3257. DOI: 10.1557/jmr.2018.320

[69] Smallman RE, Ngan AHW. Modern Physical Metallurgy. 8th ed. United States of America: Elsevier; 2014. 690 p. DOI: 10.1016/C2011-0-05565-5

[70] Hirth JP, Nix WD. An analysis of the thermodynamics of dislocation glide. Physica Status Solidi b. 1969;**35**:177. DOI: 10.1002/pssb.19690350116

[71] Hertzberg RW, Vinci RP, Hertzberg JL. Deformation and Fracture Mechanics of Engineering Materials. 5th ed. United States of America: Wiley; 2012. 800 p. ISBN-13: 978-0470527801

[72] Hsieh KT, Lin YY, Lu CH, Yang JR, Liaw PK, Kuo CL. Atomistic simulations of the face-centered-cubic-to-hexagonal-close-packed phase transformation in the equiatomic CoCrFeMnNi high entropy alloy under high compression. Computational Materials Science. 2020; **184**:109864. DOI: 10.1016/j.commatsci.2020.109864

Chapter 5

High-Entropy Superconducting Materials

Pablo Cayado and Jens Hänisch

Abstract

High-entropy alloys are a rapidly evolving field of materials research and have gained increasing attention in recent years. Characterized by their multicomponent compositions, they were originally created by mixing a multitude of metallic elements and expanded the idea of traditional alloy design, opening new paths for materials research. In particular, the discoveries of superconductivity in some of these alloys since 2014 provided a new impetus for exploring novel superconducting phenomena and materials. In fact, the concept of increasing entropy in superconducting compounds, firstly restricted to alloys or intermetallics, was soon extended to other types of superconductors enriching drastically the research in the field. The high-entropy superconductors are nowadays a matter of intense study. This chapter provides a comprehensive review of the most recent and relevant research on the different types of high-entropy superconductors. The current state of research, synthesis methods, and characterization techniques are included. This information may serve as a reference for future research on this topic and inspire further exploration and innovation in the application of these materials.

Keywords: superconductor, high-entropy alloy, high-entropy ceramic, high-entropy oxide, REBCO

1. Introduction

High-entropy alloys (HEAs) are a category of materials that has attracted attention in recent times in different fields of materials research. This new paradigm for alloy design, proposed around two decades ago [1, 2], consists of multiple species of atoms randomly intermixed, resulting in a notable level of disorder, specifically in high configurational entropy. This severe disorder appears due to the mixing of several atom species in considerable proportions on specific lattice sites.

The definition of a HEA, or, in other words, the difference between HEA and medium-entropy alloys or even the usual low-entropy alloys, is based on the number of elements present in the alloy and their proportions. An alloy is usually called HEA if it contains a mixture of at least five elements, typically in proportions between 5% and 35% [1], although some authors claim that mixtures of four elements may also be considered HEAs [3]. The entropy of such mixtures can be evaluated as:

$$\Delta S_{\text{mix}} = -R \sum_{i=1}^{n} x_i \ln x_i \qquad (1)$$

where x_i is the molar fraction of component i, n is the total number of components, and R = 8.314 J/mol K is the gas constant. An equimolar system with five components has a ΔS_{mix} value of 1.609R, which is also often used as an equivalent, stricter criterion for HEA [4]. In the following, we keep the definition of at least five components taking part in mixing while softening occasionally the criteria of composition and ΔS_{mix}.

The change in entropy lowers the Gibbs energy according to:

$$\Delta G_{\text{mix}} = \Delta H_{\text{mix}} - T\Delta S_{\text{mix}} \qquad (2)$$

with H_{mix} being the mixing enthalpy, which allows the stabilization of disordered compounds into certain singular crystal structures with substantial chemical substitutional disorder, rather than leading to the creation of multiple crystalline phases.

Among all the novel properties that emerged in HEAs, the appearance of superconductivity in some of them is one of the most interesting ones. This was seen for the first time in 2014 in the system $Ta_{34}Nb_{33}Hf_8Zr_{14}Ti_{11}$ [5] with a rather high critical temperature, T_c, of ~7.3 K [5]. This finding was the starting point of intense research into the topic as the combination of superconductivity and superior mechanical properties that HEAs typically gives them some application potential to be evaluated and utilized. Due to the large degree of disorder and hence short electron mean free paths and therefore coherence lengths, all HEA superconductors are type II. Often, their B_{c2} is measured resistively, rarely also their B_{c1} magnetically. Critical current densities are usually not reported with the exception of the $RE\mathrm{Ba_2Cu_3O_{7-x}}$ (*RE*BCO, *RE* rare earth elements) compounds. Most HEAs are fabricated by arc melting. In recent years, further preparation methods were investigated, such as solid-state reaction (SSR), spark-plasma sintering, hot-press sintering, and thin-film growth (sputtering, pulsed laser deposition, and chemical solution deposition).

Over the years, the concept of "high-entropy," that is, incorporating multiple components within the same matrix, has been expanded to various types of superconductors like BiS_2-based, Van der Waals, rock-salt-type, A15-type, and transition-metal-zirconide ($Tr\mathrm{Zr_2}$)-type superconductors. Most recently, this approach has been further extended to intermetallics as well as compounds containing non-metallic constituents like oxides, carbides, borides, nitrides, and sulfides. Particularly interesting in the field of superconductivity are the high-entropy oxides (HEO), in which the well-known *RE*BCO compounds, probably the superconductors with the largest potential from the applications point of view, are included. The research on HE *RE*BCO has started only few years ago, but it is a topic with promising perspectives. There is considerable variety in stoichiometry notation in literature.
We occasionally carefully modified it for better comparability. For the alloys, we adopted 100% notation and (near) equimolar mixes without further subscripts, that is, the number outside of the brackets represents the sum of the chemical compositions of the elements in parentheses, for instance, $(TaNb)_{67}$ refers to $Ta_{1/3}Nb_{1/3}$. Parts in 6 or 7, we note as 1/6 and 1/7 to avoid odd numbers and false accuracy. For the HE compounds, we mostly used the stoichiometry notation of the parent compounds.

1.1 Categorizing

Considering the composition and the structural and physical characteristics of the different HEAs, Sun et al. proposed a first classification by dividing them into four different groups [4]. The group of type A includes HEA superconductors made of metals from the left side of the transition-metal region of the periodic table like the different stoichiometries of the Ta-Nb-Hf-Zr-Ti system and most of the related compounds coming from elemental substitutions such as Nb-Re-Hf-Zr-Ti or Hf-Nb-Ti-V-Zr. These type A HEAs crystallize in bcc lattices and show T_c values between 4.0 K and 9.2 K and a maximum $B_{c2}(0)$ of ~11.7 T. The type B HEA superconductors are mainly composed of early transition metals (groups 3 to 7 of the periodic table), but crystallize in rather complex lattices, which could be large bcc α-Mn-type lattices or smaller bcc and hcp lattices, depending on the stoichiometry. Compounds like $(ZrNb)_{30}(MoReRu)_{70}$, $(HfTaWIr)_{70}Re_{30}$, or $(HfTaWPt)_{50}Re_{50}$ belong to this group. The T_c in this class is usually lower than for type A HEAs with values between 1.9 K and 5.7 K, and the highest known $B_{c2}(0)$ is ~7.9 T. The third group, type C HEA superconductors, crystallizes in a CsCl-type structure and is composed of early transition metals combined with late transition metals, such as the Sc-Zr-Nb-Ta-Rh-Pd system. They show T_c values between 3.9 K and 9.3 K and $B_{c2}(0)$ values up to ~10.7 T. The fourth group, type D HEA superconductors, crystallizes in the hcp structure. Examples are $Re_{56}Nb_{11}Ti_{11}Zr_{11}Hf_{11}$, $(MoReRuRh)_{95}Ti_5$, and $Nb_{10+2x}Mo_{35-x}Ru_{35-x}Rh_{10}Pd_{10}$. This last alloy shows the highest T_c of this class, with a value of 6.19 K for $x = 2.5$. With the appearance of more and more different types of superconducting high-entropy compounds, this classification was extended by naming them simply according to their crystal structure.

One parameter that may have a great influence on the stability of HEAs is the valence electron count (VEC), which is better expressed as the number of valence electrons per atom in the particular case of HEAs. The VEC reflects the total density of states at the Fermi level and plays a crucial role in determining the T_c of superconductors that follow the Bardeen-Cooper-Schriefer (BCS) theory. In fact, a robust correlation between VEC and T_c is evident in binary and ternary superconducting transition-metal alloys, commonly known as the Matthias rule [6]. In the case of HEAs, there is also a relationship between the VEC and T_c similar to the Matthias rule. In **Figure 1**, the dependence of T_c on the VEC is shown for different types of superconducting HEAs. The type A HEAs that crystallize in bcc structure appear in the VEC range 4.2–4.8, reaching a maximum T_c for a VEC of 4.7. In the case of type C HEAs, with CsCl-type structures, superconductivity emerges in the VEC range of 5.9–6.3, with a maximum T_c at a VEC of ~5.9. These two types of HEAs have similar behavior to the simpler binary alloys with the characteristic two-peak character seen for the near VECs of 4.7 and 5.9. However, the type B HEAs with a more complex α-Mn structure do not follow this general trend. Although the bcc and CsCl crystal structures are closely related to the α-Mn structure, their atomic arrangements differ in a remarkable way, making their stabilities to be limited to narrow ranges of VEC. The type D HEAs, the ones that show hcp structure, show superconducting behavior for VEC values similar to the type C ones. Finally, it is possible to add to this analysis the more recent HEAs. The A15-type HEAs show similar behavior as the binary A15 superconductors that follow the Matthias rule, with VEC falling between 4.55 and 4.85. Also, the σ-phase HEAs resemble their relatives, the binary σ-phase superconductors whose VEC values extend from 6.07 to 6.84 with a T_c maximum at VEC ~ 6.5 in accordance with the Matthias rule [6]. The VECs of the σ-phase HEAs are located at

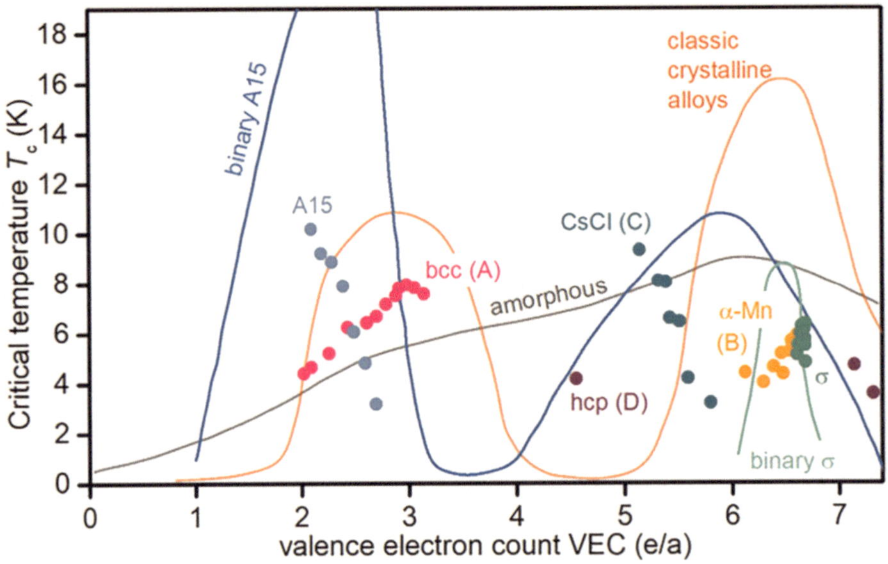

Figure 1.
Dependence of the critical temperature T_c *on the valence electron count (VEC) for different types of high-entropy superconductors. Data extracted from Refs. [4, 7, 8].*

the electron-rich side of this region and are higher than those of most HEA superconductors. Note that the VEC ranges of the σ phase and the α-Mn type HEA superconductors are very close. These dependencies show that the VEC serves as a crucial tool for predicting potential HEA superconductors and can be useful for understanding the superconducting properties of the new systems.

Below, the state of the art of each of these HE superconductor classes will be reported. Special attention will be paid to the superconducting properties of each compound and their prospects for further development.

2. Superconducting high-entropy alloys

2.1 The Ta-Nb-Hf-Zr-Ti system

Superconductivity in high-entropy alloys was first discovered by Koželj et al. in 2014 for a particular stoichiometry of the Ta-Nb-Hf-Zr-Ti system [5]. Investigations on this system had already started years prior due to its superior structural and mechanical properties [9]. Varying the original equimolar ratio of the elements led to the $Ta_{34}Nb_{33}Hf_8Zr_{14}Ti_{11}$ compound that appeared to be superconducting below a critical temperature of ~7.3 K. The original compound was a polycrystalline sample with grains in the range of 2–300 μm fabricated by arc melting. It crystallized in a highly distorted body-centered cubic (bcc) structure (a = 3.36 Å), **Figure 2a**, due to the difference in the atomic radii of the constituent elements. Apart from its ~7.3 K T_c, it shows an upper critical field $B_{c2}(0) \sim 8.2$ T, a lower critical field $B_{c1} \sim 32$ mT, and an energy gap opening of ~2.2 meV. Moreover, first principle calculations showed that it could be classified as a strong electron-phonon coupling superconductor with a coupling constant $\lambda = 1.16$ [11].

This discovery soon caught the attention of several groups, who started to investigate this system, preparing the samples, as in the original work, by an arc melting

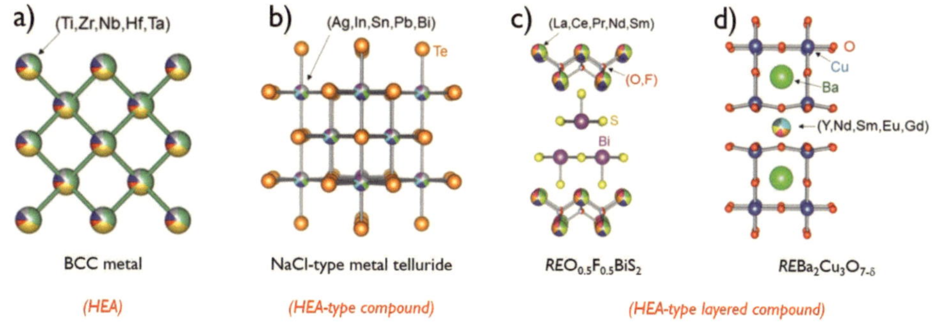

Figure 2.
Crystal structures of several high-entropy superconductors. Reproduced with permission from [10].

method as summarized in **Table 1**. A stoichiometry series, studied a couple of years later by von Rohr et al., was $[TaNb]_{100-x}(HfZrTi)_x$, where x took 12 different values from 20 to 84 [13]. This study revealed T_c variations between 4.49 K and 7.92 K for the different stoichiometries, with the largest upper critical field of ~11.67 T for the sample with x = 0.5. A particular stoichiometry of this compound, $(TaNb)_{67}(HfZrTi)_{33}$, was further studied by Guo et al. [14] at high pressures, revealing an exceptionally resilient zero-resistance superconductivity under pressures up to 190.6 GPa. In their study, the authors observed that initially, the T_c of this compound rose from 7.7 K at ambient pressure to 10 K at ~60 GPa, followed by a slight, gradual decrease to 9 K at 190.6 GPa, which is close to the pressure in the outer core of the Earth. The authors proposed therefore that this compound is a potential candidate for new applications under extreme pressure conditions. Similar results confirming the robustness of the superconductivity in the Ta-Nb-Hf-Zr-Ti system were obtained by Vrtnik et al. [12], who prepared samples with different numbers of elements (4 and 5) as well as differing stoichiometry (equimolar and off-equimolar) by arc melting and subjected them to different thermal treatments. Their results show that the samples are superconducting independently of the thermal treatment and the resulting microstructure. In addition, Jasiewicz et al. [19] further investigated the pressure effects on the electronic structure, electron-phonon interaction, and superconductivity of $(TaNb)_{67}(HfZrTi)_{33}$ from 0 to 100 GPa. Their investigation confirmed the previously reported results, observing the same T_c trend with pressure. However, they could also prove that this evolution of T_c with pressure can be well explained in the framework of the classical electron-phonon mechanism. A certain variant also investigated was $Ta_{1/6}Nb_{2/6}Hf_{1/6}Zr_{1/6}Ti_{1/6}$ by Kim et al. [16]. With a T_c = 7.85 K, it shows a $B_{c1}(0)$ ~ 23 mT and a $B_{c2}(0)$ ~ 12.05 T, a higher value than the one found for the $(TaNb)_{50}(HfZrTi)_{50}$ commented before. This work was the first one to report about critical current density and pinning mechanisms in this type of systems. The authors reported a critical current density (J_c) of ~10.7 kA/cm^2 at 2 K in self-field and found that the pinning mechanism agreed with the double exponential model, which was first introduced for MgB_2 by Horvat et al. [20], indicating that there were two types of pinning mechanism active. Kim et al. [17] investigated recently the effect of vacuum annealing at several temperatures and at different times on the superconducting and mechanical properties of this compound. Even though T_c and B_{c2} slightly decreased upon annealing at medium temperatures (550°C), it had a positive effect on both J_c and Vickers hardness (HV_{IT}). Record values

Nominal composition	T_c (K)	$B_{c1}(0)$ (mT)	$B_{c2}(0)$ (T)	Reference
$Ta_{34}Nb_{33}Hf_8Zr_{14}Ti_{11}$	7.3	32	8.2	[5]
$Ta_{34}Nb_{33}Hf_8Zr_{14}Ti_{11}$	7.8	—	8.15	[12]
$[TaNb]_{70}(ZrHfTi)_{30}$	8.03	—	6.67	[13]
$[TaNb]_{67}(ZrHfTi)_{33}$	7.75	—	7.75	[13]
$[TaNb]_{67}(ZrHfTi)_{33}$	7.7 (1 atm)	—	—	[14]
$[TaNb]_{67}(ZrHfTi)_{33}$	10 (60 GPa)	—	—	[14]
$[TaNb]_{67}(ZrHfTi)_{33}$	9 (190.6 GPa)	—	—	[14]
$[TaNb]_{60}(ZrHfTi)_{40}$	7.56	—	8.43	[13]
$[TaNb]_{50}(ZrHfTi)_{50}$	6.46	—	11.67	[13]
$[TaNb]_{16}(ZrHfTi)_{84}$	4.52	—	9.02	[13]
$Ta_{20}Nb_{21}Hf_{20}Zr_{20}Ti_{19}$	6.9	—	10.45	[12]
$Ta_{20}Nb_{20}Hf_{20}Zr_{20}Ti_{20}$	7.12			[15]
$Ta_{22}Nb_{24}Hf_{21}Zr_{23}Ti_{10}$	6.2	—	—	[12]
$Ta_{1/6}Nb_{2/6}Hf_{1/6}Zr_{1/6}Ti_{1/6}$	7.85	23	12.05	[16]
$Ta_{1/6}Nb_{2/6}Hf_{1/6}Zr_{1/6}Ti_{1/6}$	8.1		11.0	[17]
$Ta_{11}Nb_{34}Hf_8Zr_{14}Ti_{33}$	7.5	15	12.2	[18]

Table 1.
Selected properties of superconducting Ta-Nb-Hf-Zr-Ti HEAs, all prepared by arc melting.

of $J_c(0)$ = 520 kA/cm^2 and HV_{IT} = 528 were reported, an increase of 1860% and 32%, respectively. The best samples of Refs. [16, 17], **Table 1**, investigated are (also) included in **Table 2** for comparison of J_c. The already high value of $B_{c2}(0)$ found in [16] was recently overcome in the $Nb_{34}Ti_{33}Zr_{14}Ta_{11}Hf_8$ alloy with $B_{c2}(0) \sim 12.2$ T, although it has a lower T_c = 7.5 [18]. For a summary of the Ta-Nb-Hf-Zr-Ti HEAs, see **Table 1**.

2.2 Substitutions in Ta-Nb-Hf-Zr-Ti

The natural evolution of the field led to the study of new compositions and mixtures of elements, first by only replacing one of the elements. These compounds, summarized in **Table 3**, crystallize in the bcc structure and belong to type A too. An alternative soon investigated by Marik et al. [27] was the equimolar Nb-Re-Hf-Zr-Ti HEA, the result of substituting Re for Ta in the original system. Prepared by arc melting too, it was among the first equimolar HEA superconductors. With $T_c \sim 5.3$ K, it shows an upper critical field $B_{c2}(0)$ of 8.88 T, a value close to the BCS weak coupling Pauli limit, in contrast to other compounds of the Ta-Nb-Hf-Zr-Ti system [17]. The same elements in different stoichiometry were employed by Motla et al. [28] in the synthesis of $Nb_{60}Re_{10}Zr_{10}Hf_{10}Ti_{10}$. This variant shows a T_c = 5.7 K, and it was studied in detail by means of different electronic characterization techniques, including muon spin relaxation and rotation, finding electronic properties similar to conventional binary and ternary superconducting compounds, that is, s-wave symmetry of the order parameter and phonon-mediated pairing [18]. Ta was also substituted by Mo in the equimolar HfMoNbTiZr, resulting in T_c = 4.1 K and $B_{c2}(0)$ = 5.55 T [30]. Re, on

Nominal composition	Synthesis method	T_c (K)	$B_{c2}(0)$ (T)	J_c (kA/cm^2)	Reference
$Ta_{1/6}Nb_{2/6}Hf_{1/6}Zr_{1/6}Ti_{1/6}$	Arc melting	7.85	12.05	10.7 (2 K, sf)	[16]
$Ta_{1/6}Nb_{2/6}Hf_{1/6}Zr_{1/6}Ti_{1/6}$	Arc melting	8.1	11	520 (4 K, sf)	[17]
$Ta_{1/6}Nb_{2/6}Hf_{1/6}Zr_{1/6}Ti_{1/6}$	SPS	7.80	10.50	30.5 (2 K, 0.01 T) 73.2 (4 K, 0.01 T)	[21]
$Ta_{1/6}Nb_{2/6}Hf_{1/6}Zr_{1/6}Ti_{1/6}$	Hot-press sintering	7.90	—	—	[22]
$Ta_{1/6}Nb_{2/6}Hf_{1/6}Zr_{1/6}Ti_{1/6}$	PLD	7.28	12.10	1.8 (4.2 K, sf)	[23]
$(TaNb)_{12}(HfZrTi)_{88}$	Sputtering	2.77	6.15	—	[24]
$(TaNb)_{24}(HfZrTi)_{76}$	Sputtering	4.61	10.44	—	[24]
$(TaNb)_{35}(HfZrTi)_{65}$	Sputtering	5.60	11.05	—	[24]
$(TaNb)_{46}(HfZrTi)_{54}$	Sputtering	6.14	9.93	—	[24]
$(TaNb)_{57}(HfZrTi)_{43}$	Sputtering	6.76	8.77	—	[24]
$(TaNb)_{67}(HfZrTi)_{33}$	Sputtering	6.43	7.05	—	[24]
$(TaNb)_{79}(HfZrTi)_{21}$	Sputtering	6.33	5.78	—	[24]
$(TaNb)_{87}(HfZrTi)_{13}$	Sputtering	6.02	4.29	—	[24]
$(TaNb)_{96}(HfZrTi)_{04}$	Sputtering	5.57	2.95	—	[24]
$(TaNb)_{70}(HfZrTi)_{30}$ (600 nm)	Sputtering	5.31	5.80	—	[25]
$(TaNb)_{70}(HfZrTi)_{30}$ (100 nm)	Sputtering	2.69	3.20	—	[25]
$(TaNbZrTi)_{90}W_{10}$	Sputtering	6.20	—	—	[26]
$(TaNbZrTi)_{90}V_{10}$	Sputtering	6.20	—	—	[26]

Table 2.
Selected properties of superconducting Ta-Nb-Hf-Zr-Ti HEAs prepared by methods different from the conventional arc melting.

the other hand, was also used by Hattori et al. [29] to substitute not the Ta but the Zr, leading to the Ti-Hf-Nb-Ta-Re system. Different stoichiometries with VEC values ranging from 4.6 to 5.0 were investigated, finding T_c values between 3.25 K and 4.38 K and $B_{c2}(0)$ values between 7.90 T and 5.91 T, with the maximum T_c occurring for a VEC of 4.7 and the maximum $B_{c2}(0)$ for a VEC of 4.6 [27]. The Zr was also substituted, in this case by Mo by Zeng et al. [31], to prepare equimolar TiHfNbTaMo samples that showed a T_c = 3.42 K and $B_{c2}(0)$ = 3.95 T [28]. The Zr was also replaced by V in the equimolar HfNbTaTiV HEA [34]. The detailed characterization determined T_c = 5.0 K, B_{c1} = 19 mT and $B_{c2}(0)$ = 6.63 T [30]. Another substitution, V for Ta, led to $Hf_{21}Nb_{25}Ti_{15}V_{15}Zr_{24}$, which shows a T_c value of 5.3 K [33]. In the case of $[TaNb]_{31}(TiUHf)_{69}$, the substitution was U for Zr [35]. This was the first HEA containing an *f*-electron element, and it extended the HEA research to actinide elements, opening the possibility of leveraging HEAs as functional waste forms for a variety of radioisotopes. This system shows phonon-mediated superconductivity with $T_c \sim 3.2$K and $B_{c2}(0) \sim 6.4$T [31].

Other elemental substitutions of two or more elements in the Ta-Nb-Hf-Zr-Ti system were studied by von Rohr et al. [36] with systematic isoelectronic replacements, using Mo-Y, Mo-Sc, and Cr-Sc mixtures and also adding Al to the initial

Nominal composition	T_c (K)	$B_{c1}(0)$ (mT)	$B_{c2}(0)$ (T)	Reference
$\mathbf{Re_{20}}Nb_{20}Hf_{20}Zr_{20}Ti_{20}$	5.3	33	8.88	[27]
$\mathbf{Re_{10}}Nb_{60}Hf_{10}Zr_{10}Ti_{10}$	5.7	7.95	6.31	[28]
$Ta_5Nb_{25}Hf_{25}\mathbf{Re_{10}}Ti_{35}$	3.95	4.87	7.90	[29]
$Ta_5Nb_{35}Hf_{20}\mathbf{Re_{10}}Ti_{30}$	4.38	6.60	6.94	[29]
$Ta_{15}Nb_{35}Hf_{15}\mathbf{Re_{10}}Ti_{25}$	4.10	4.56	6.65	[29]
$Ta_{25}Nb_{35}Hf_{10}\mathbf{Re_{10}}Ti_{20}$	3.62	7.15	5.85	[29]
$Ta_{35}Nb_{35}Hf_5\mathbf{Re_{10}}Ti_{15}$	3.25	6.87	5.91	[29]
$\mathbf{Mo_{20}}Nb_{20}Hf_{20}Zr_{20}Ti_{20}$	4.1	5.2	5.82	[30]
$Ta_{20}Nb_{20}Hf_{20}\mathbf{Mo_{20}}Ti_{20}$	3.42	22.8	3.95	[31]
$(TaNb)_{0.67}(Hf\mathbf{MoW})_{0.33}$	4.3	—	1.45	[32]
$\mathbf{V_{15}}Nb_{25}Hf_{21}Zr_{24}Ti_{15}$	5.30	—	—	[33]
$Ta_{20}Nb_{20}Hf_{20}\mathbf{V_{20}}Ti_{20}$	4.93	—	6.63	[34]
$[TaNb]_{31}(Hf\mathbf{U}Ti)_{69}$	3.20	—	6.40	[35]
$[\{\mathbf{Sc_{33}Cr_{67}}\}Nb]_{67}(HfZrTi)_{33}$	5.60	—	—	[36]
$[Ta\{\mathbf{Sc_{33}Cr_{67}}\}]_{67}(HfZrTi)_{33}$	4.40	—	—	[36]
$[TaNb]_{67}(\{\mathbf{Sc_{67}Cr_{33}}\}ZrTi)_{33}$	7.50	—	—	[36]
$[TaNb]_{67}(Hf\{\mathbf{Sc_{67}Cr_{33}}\}Ti)_{33}$	7.40	—	—	[36]
$[TaNb]_{67}(HfZr\{\mathbf{Sc_{67}Cr_{33}}\})_{33}$	7.60	—	—	[36]
$[\{\mathbf{Y_{33}Mo_{67}}\}Nb]_{67}(HfZrTi)_{33}$	4.70	—	—	[36]
$[Ta\{\mathbf{Y_{33}Mo_{67}}\}]_{67}(HfZrTi)_{33}$	3.50	—	—	[36]
$[TaNb]_{67}(\{\mathbf{Y_{67}Mo_{33}}\}ZrTi)_{33}$	7.60	—	—	[36]
$[TaNb]_{67}(Hf\{\mathbf{Y_{67}Mo_{33}}\}Ti)_{33}$	6.70	—	—	[36]
$[TaNb]_{67}(HfZr\{\mathbf{Y_{67}Mo_{33}}\})_{33}$	7.50	—	—	[36]
$[\{\mathbf{Sc_{33}Mo_{67}}\}Nb]_{67}(HfZrTi)_{33}$	4.40	—	—	[36]
$[Ta\{\mathbf{Sc_{33}Mo_{67}}\}]_{67}(HfZrTi)_{33}$	2.90	—	—	[36]
$[TaNb]_{67}(\{\mathbf{Sc_{67}Mo_{33}}\}ZrTi)_{33}$	7.50	—	—	[36]
$[TaNb]_{67}(Hf\{\mathbf{Sc_{67}Mo_{33}}\}Ti)_{33}$	6.60	—	—	[36]
$[TaNb]_{67}(HfZr\{\mathbf{Sc_{67}Mo_{33}}\})_{33}$	7.50	—	—	[36]
$Ta_{20}Nb_{20}\mathbf{Fe_{20}}Zr_{20}Ti_{20}$	6.87	—	—	[15]
$Ta_{20}Nb_{20}\mathbf{Ge_{20}}Zr_{20}Ti_{20}$	8.46	—	—	[15]
$Ta_{1/6}Nb_{1/6}Hf_{1/6}Zr_{1/6}Ti_{1/6}\mathbf{V_{1/6}}$	5.09	—	—	[15]
$Ta_{1/6}Nb_{1/6}\mathbf{Ge_{1/6}V_{1/6}}Zr_{1/6}Ti_{1/6}$	8.40	—	—	[15]
$Ta_{1/6}Nb_{1/6}\mathbf{Si_{1/6}V_{1/6}}Zr_{1/6}Ti_{1/6}$	4.29	—	—	[15]
$Ta_{1/6}Nb_{1/6}\mathbf{Si_{1/6}Ge_{1/6}}Zr_{1/6}Ti_{1/6}$	7.40	—	—	[15]
$Ta_{20}Nb_{20}Hf_{20}\mathbf{Sc_{20}}Ti_{20}$	6.60	—	13.10	[37]
$Ta_{20}Nb_{20}Hf_{20}Zr_{20}\mathbf{Sc_{20}}$	7.70	—	12.40	[37]
$Ta_{20}Nb_{20}\mathbf{Sc_{20}}Zr_{20}Ti_{20}$	7.90	—	19.30	[37]
$Ta_{1/6}Nb_{1/6}Hf_{1/6}Zr_{1/6}Ti_{1/6}\mathbf{Sc_{1/6}}$	7.20	—	14.10	[37]

Nominal composition	T_c (K)	$B_{c1}(0)$ (mT)	$B_{c2}(0)$ (T)	Reference
$\mathbf{Al_{15}}Nb_{24}\mathbf{V_5}Zr_{26}Ti_{40}$	5.36	—	—	[38]
$\mathbf{Al_{15}}Nb_{14}\mathbf{V_5}Zr_{41}Ti_{35}$	4.73	—	—	[38]
$\mathbf{Al_{15}}Nb_{24}\mathbf{V_5}Zr_{31}Ti_{35}$	5.45	—	—	[38]
$\mathbf{Al_{15}}Nb_{34}\mathbf{V_5}Zr_{21}Ti_{35}$	6.47	—	—	[38]
$\mathbf{Al_{15}}Nb_{44}\mathbf{V_5}Zr_{11}Ti_{35}$	6.75	—	—	[38]
$\mathbf{Al_{15}}Nb_{14}\mathbf{V_5}Zr_{31}Ti_{45}$	4.96	—	—	[38]
$\mathbf{Al_{15}}Nb_{34}\mathbf{V_5}Zr_{31}Ti_{25}$	6.18	—	—	[38]
$\mathbf{Al_{15}}Nb_{44}\mathbf{V_5}Zr_{31}Ti_{15}$	6.98	—	—	[38]

Table 3.
Selected properties of superconducting type A Ta-Nb-Hf-Zr-Ti-based HEAs with different substitutions (bold). All samples are prepared by arc melting.

$[TaNb]_{67}(HfZrTi)_{33}$ compound ($[TaNb]_{67}(HfZrTi)_{33}Al_x$). This study concluded that T_c strongly depends on the elemental composition of the alloy. This change is clearly visible when substituting Nb and also Ta, observing a 60% decay in T_c of $[Ta\{Sc_{33}Mo_{67}\}]_{67}(HfZrTi)_{33}$. The replacement of the other three metals does not change T_c so drastically, which remains almost unchanged for the replacement of Hf and Ti and being ~1 K lower for the replacement of Zr. In the case of the Al addition, an improvement of the crystallinity of the compounds was realized until $x = 40$, when the stabilization of a simple bcc lattice breaks down and the material starts to crystallize in the β-uranium structure, resulting in a suppression of superconductivity below 1.8 K [34]. Similar conclusions can be extracted from the work of Wu et al. [15], who added Fe, Ge, Hf, Si, and/or V instead of Hf to the original NbTaTiZr-based matrix. The analysis of the physical properties of the prepared compounds, including up to seven elements, revealed that the superconducting behavior varies with the composition distribution of Nb+Ta and the degree of lattice distortion generated by the change in the total number of elements. As the number of elements increases from four to seven, the Nb+Ta content reduces, causing a decrease in T_c [20]. The addition of Sc to thc NbTaTiZr-based matrix generates a two-phase structure, with one phase being bcc, very similar to the Hf-Nb-Ta-Ti-Zr parent system and the other one hcp, which absorbs practically all Sc. Superconductivity emerges in the bcc phase with $T_c = 7.2$ K and $B_{c2}(0) = 14.1$ T, very similar to the Hf-Nb-Ta-Ti-Zr parent alloys [37]. Another studied system was Al-Nb-Ti-V-Zr. Harayama et al. [39] prepared $Al_5Nb_xTi_{35}V_5Zr_{55-x}$ and $Al_5Nb_xTi_{59-x}V_5Zr_{31}$ samples, including the particular composition $Al_5Nb_{24}Ti_{40}V_5Zr_{26}$, which have an interesting gum-metal-like behavior characterized by being superelastic and having a low Young's modulus [35]. The T_c values of these compounds vary from 5.36 K to 7.01 K, being the maximum for the $Al_5Nb_{44}Ti_{15}V_5Zr_{31}$ sample, following the T_c dependence on the VEC typical for the bcc-type HEA superconductors of type A [38]. The $(TaNb)_{67}(HfMoW)_{33}$ HEA was also investigated in detail. This HEA exhibits a metallic behavior at high temperature, while at $T_c \sim 4.3$ K, it becomes superconducting with an upper critical field $B_{c2}(0) \sim 1.45$ T [32].

2.3 Preparation methods

Until recent times, superconducting HEAs were generally prepared by arc melting. In the search for new features, alternative preparation methods were started, summarized in **Table 2**. Some of them, like Spark-Plasma Sintering (SPS) employed to synthesize $Ta_{1/6}Nb_{2/6}Hf_{1/6}Zr_{1/6}Ti_{1/6}$ samples, serve to obtain bulk samples similar to those prepared by arc melting. The T_c = 7.8 K and $B_{c2}(0)$ = 10.5 T of these samples are similar to the one obtained in the samples by arc melting. However, the difference in the preparation method led to a very different pinning landscape. While the behavior of arc-melted samples is described by the surface pinning model, the SPS sintered sample includes also point pinning defects that sum up to the contribution of the surface pinning. This leads to a J_c of 30.5 kA/cm^2 at 2 K and 0.01 T and 73.2 kA/cm^2 at 4 K and 0.01 T, values that are 286% and 687% larger, respectively, to those obtained in the arc-melted sample [21]. Samples with exactly the same composition were also prepared by hot-press sintering. Also these samples show a similar T_c = 7.9 K as samples prepared by arc melting [22].

However, the most remarkable and interesting advances in alternative preparation methods are observed in the HEA films. The first work about HEA films was published by Zhang et al. and was about the preparation of $(TaNb)_{100-x}(HfZrTi)_x$ films on Si substrates by DC magnetron co-sputtering at room temperature [24]. These films, with different stoichiometries and thicknesses (between 600 and 950 nm), crystallize, like their bulk relative, into a bcc structure. All the films are superconducting with T_c values in the range of 2.77–6.76 K. The largest T_c is found for a x = 43, and it is around 1 K lower than the bulk sample with the same composition due to, probably, the confined-size effect. The largest $B_{c2}(0)$ = 11.05 T is found, however, for x = 65, which even exceeds the Pauli paramagnetic limit [24]. Films of a different stoichiometry of the same system, the $(TaNb)_{70}(HfZrTi)_{30}$, were also deposited by magnetron sputtering to investigate how the thickness and the external hydrostatic pressure affect the superconducting properties. It was proved that both T_c and $B_{c2}(0)$ decay with decreasing thickness. The T_c values decrease from the original 7.8 K for the bulk to 5.31 K for the 600 nm film and 2.69 K for the 100 nm film. Films with a thickness of 30 nm are not superconducting. The same trend is observed for $B_{c2}(0)$, with values of 6.7, 5.8, and 3.2 T, for bulk, 600 nm and 100 nm, respectively. The application of hydrostatic pressure up to 3.3 GPa has different consequences depending on the thickness of the film. In the case of the 600 nm film, T_c decreases with the pressure, which is different from the behavior observed on the bulk samples. However, no significant T_c change was observed when pressure was applied to the 100 nm film. This could indicate, as expected, that thinner films increase the disorder due to amorphization [25]. The same deposition technique was used in the work of Shu et al. to deposit $(TaNbZrTi)_{100-x}W_x$ and $(TaNbZrTi)_{100-x}V_x$ superconducting films. In this work, the influence of alloying the TaNbZrTi-based system with W and V and the deposition temperature on the phase formation and superconducting behavior of the films was investigated. The results show that all the films with different values of x show a bcc structure similar to the original TaNbZrTi system. In terms of T_c values, the highest one, 8.0 K, was reached for the original TaNbZrTi system. Then, for both alloying elements, T_c dropped monotonically as a function of x, almost at the same rate. When increasing the deposition temperature from room temperature to 400°C, T_c increases while the compositions remain unchanged. Moreover, the experimental observations led to the conclusion that the T_c values may also relate to the atomic radii difference and electronegativity difference of involved elements beyond the valence electron number [26]. A very similar study was done by Zhang et al. with $(TaNb)_{100-x}(HfZrTi)_xMo_y$ films. Films with different stoichiometries of $\sim$300 nm were deposited on Si substrates by magnetron sputtering. The films show again a bcc structure for low x values. For near-equimolar composition, the crystalline HEA grains transform into amorphous aggregations with a size of a few nanometers, forming a crystal/glass nanocomposite. This particular microstructure causes a broadening of the normal-to-superconducting transition and suppresses the zero-resistivity critical temperature to a lower constant value of approximately 2.9 K [40]. A different technique, namely pulsed laser deposition (PLD), was employed by Jung et al. to deposit $Ta_{1/6}Nb_{2/6}Hf_{1/6}Zr_{1/6}Ti_{1/6}$ films. This HEA, studied in several works in bulk samples, was deposited on Al_2O_3 substrates at different temperatures ranging from 270 to 620°C. These films, which crystallize as expected in bcc structure, show T_c values between ~5.4 and 7.3 K, being the maximum value for deposition temperature of 520°C. Also, for the same deposition temperature, the largest $B_{c2}(0)$ = 12.1 T is observed. All films showed significantly larger J_c values than the bulk sample, while a self-field value above 1 MA/cm^2 (~ 1.8 MA/cm^2) at 4.2 K was measured for the film deposited at 520°C. The J_c values were maintained above 0.1 MA/cm^2 under magnetic fields of approximately 3.4 and 2.0T at 2.0 and 4.2 K, respectively, which is approximately eight times larger than in the bulk sample at the same conditions. Another remarkable result of this work is the effect of ion irradiation on the properties of the films. It was demonstrated that the HEA films are over 1000 times more robust to ion irradiation than other superconducting materials, such as Nb_3Sn, MgB_2, Fe-based superconductors, or *RE*BCO. This means that the T_c of this HEA remains stable for much larger ion doses [23].

2.4 Other alloys

Rather different from the Ta-Nb-Hf-Zr-Ti system are the pentanary $(ScZrNb)_{1-x}[RhPd]_x$ and hexanary $(ScZrNbTa)_{100-x}[RhPd]_x$ systems reported by Stolze et al. [41]. They have investigated several mixtures of five and six elements showing a CsCl-type lattice, which is an ordered bcc structure, but had not been found in the original Ta-Nb-Hf-Zr-Ti system. It was also observed that T_c increases gradually with decreasing valence electron count (VEC) within a solution series. The highest T_c and $B_{c2}(0)$ are observed in the particular case of $(ScZrNb)_{65}[RhPd]_{35}$ with values of ~9.3 K and 10.7 T, respectively. Another important parameter that was observed in this case was the chemical composition, where the presence of niobium has a positive impact on the superconductivity. In any case, also samples without any niobium were superconducting proving that superconductivity is an intrinsic feature of the bulk material and not just an average of the constituents' elemental properties [26]. The elastic properties of such a hexanary HEA superconductor, $(ScZrNbTa)_{68.5}[RhPd]_{31.5}$, was investigated by Pan et al. [42]. Its Young's modulus E = 121.0 GPa and shear modulus of G = 44.9 GPa are around 10% larger than for Nb-Ti of similar T_c and ductility, which is promising for applications.

The authors of Ref. [41] reported also on three other new pentanary systems: $(ZrNb)_{1-x}[MoReRu]_x$, $(HfTaWIr)_{100-x}[Re]_x$, and $(HfTaWPt)_{100-x}[Re]_x$, which crystallize in a bcc α-Mn-type structure firstly observed in superconducting HEAs here [43]. The last two systems reported in this work are the first purely 5d-metal-based superconducting HEAs. Several stoichiometries were tested, showing T_c values between 2.1 K and 6.3 K. Just as in the case of type A Re-Nb-Hf-Zr-Ti, the presence of Re is important to understand the behavior of these systems. Here, Re plays a key role in the stabilization of this structure type [23].
Liu et al. [44] observed the same α-Mn-type structure in $Nb_{25}Mo_{5+x}Re_{35}Ru_{25-x}Rh_{10}$ $(0 \leq x \leq 10)$, of which $Nb_{25}Mo_{10}Re_{35}Ru_{20}Rh_{10}$ shows the largest T_c = 5.1 K and $B_{c2}(0)$ = 8.3 T [41].

The general tendency of the superconducting HEAs to crystallize in cubic structures was not followed by the discovery of the $Re_{56}Nb_{11}Ti_{11}Zr_{11}Hf_{11}$ alloy by Marik et al. [45], the first hexagonal superconducting HEA. These HEAs with hexagonal-close-packed (hcp) structure are in general rather interesting from the practical point of view because of their high hardness. $Re_{56}Nb_{11}Ti_{11}Zr_{11}Hf_{11}$ has a T_c = 4.4 K, $B_{c1}(0)$ = 2.3 mT, and $B_{c2}(0)$ = 3.6 T [42]. Other superconducting HEAs crystallizing in hcp structure are, for example, $(MoReRuRh)_{95}Ti_5$ and $(MoReRuRh)_{90}Ti_{10}$ with T_c = 3.6 K and 4.7 K, respectively [46], or $Nb_{10+2x}Mo_{35-x}Ru_{35-x}Rh_{10}Pd_{10}$ $(0 \leq x \leq 5)$ with a maximum T_c = 6.19 K for x = 2.5 [47]. A variant of the archetypical hcp structure is found in transition-metal antimonides $M_{1-x}Pt_xSb$ (M = equimolar mix of Ru, Rh, Pd, and Ir) that crystallize in a pseudo-hexagonal NiAs-type crystal structure [48]. The synthesis for various Pt contents, x, and heat treatment conditions lead to entropy-stabilized transition-metal monoantimonides. Superconductivity was only found for a particular composition, x = 0.2, for which a superconducting transition at T_c = 2.15 was observed. This T_c is similar to the highest values previously reported for transition-metal monoantimonides [45].

A transformation from the original hcp structure to a new face-centered cubic (fcc) one was observed in $(MoReRu)_{(1-2x)/3}(PdPt)_x$ and carbon-added $(MoReRu)_{(1-2x)/3}(PdPt)_xC_y$ HEAs by Zhu et al. [49]. The $(MoReRu)_{(1-2x)/3}(PdPt)_x$ alloys were found to crystallize in an hcp structure for $0.042 \leq x \leq 0.167$ and in an fcc structure for x = 0.333. Their VEC values are somewhat outside the expected ranges, between 7.0 and 8.0. The addition of a certain amount of C that depends on x causes also a transformation to the fcc structure. As in the C-free case, the structural transformation is likely driven by the combined effects of VEC, lattice distortion, and increase in entropy. The T_c values of the hcp phases range from 1.75 K to 8.17 K, while the ones of the fcc structures are always below 2.8 K [46].

A different crystal structure is observed in the $Ta_5[Mo_{35-x}W_{5+x}]Re_{35}Ru_{20}$ and $[Ta_{5+y}Mo_{35-y}]W_5Re_{35}Ru_{20}$ HEAs [8]. In this case, the HEAs crystallize into tetragonal sigma (σ) phases. These systems exhibit fully gapped bulk superconductivity with T_c values between 5.27 K and 6.29 K, reaching the maximum at $x = y = 0$ [47]. The same type of structure was observed in $Ta_{10}Mo_{35-x}Cr_xRe_{35}Ru_{20}$ $(5 \leq x \leq 13)$ [50], highlighted by the presence of Cr, which is not commonly found in superconducting materials. These HEAs exhibit T_c values up to 4.79 K and show a paramagnetic behavior in the normal state due to the magnetic moment of Cr [48].

A transformation from the σ-phase into a new structure happens in $Ta_{10}Mo_5W_{30}Re_{35}Ru_{20}$ upon adding C [51]. The $Ta_{10}Mo_5W_{30}Re_{35}Ru_{20}C_x$ HEA crystallizes in the centrosymmetric σ-type structure for $0 \leq x \leq 2$ but transforms into the noncentrosymmetric β-Mn structure for $16 \leq x \leq 20$. The HEA with the β-Mn structure shows T_c = 5.34 K, which is larger than for the original σ-type structure. The $B_{c2}(0)$ = 9.3 T is close to the Pauli paramagnetic limit. This phase transformation is likely caused by the decrease in VEC and the increase in lattice distortion and mixing entropy [49]. The same β-Mn structure is found in the $Cr_{5+x}Mo_{35-x}W_{12}Re_{35}Ru_{13}C_{20}$ HEA $0 \leq x \leq 9$ [52]. By increasing x, T_c was observed to decrease from 5.49 K to 3.35 K due to the magnetic pair breaking caused by Cr moments. For all the different stoichiometries, $B_{c2}(0)$ is comparable to the Pauli paramagnetic limit $B_P(0) = 1.86T_c$, having a maximum value of 9.7 T [50].

Based on the relationship between the VEC and T_c values for HEAs and other binary or ternary superconducting alloys, new combinations of elements and stoichiometries were tested approaching the VEC values where superconductivity appears. As a result, several new HEA systems have been discovered. One example is the new boron-based HEA superconductor $Mo_{11}W_{11}V_{11}Re_{34}B_{33}$ (MWVRB) [53]. With a VEC = 5.24, this compound crystallized in the tetragonal $CuAl_2$ crystal structure and showed a T_c = 4.0 K and a $B_{c2}(0)$ = 7.3 T [51]. For a summary of these HEAs different from TaNbHfZrTi basis, see **Table 4**.

Structure	Nominal composition	T_c (K)	$B_{c2}(0)$ (T)	Reference
CsCl	$(ScZrNb)_{60}[RhPd]_{40}$	5.20	2.10	[41]
	$(ScZrNb)_{62}[RhPd]_{38}$	9.20	8.90	[41]
	$(ScZrNb)_{63}[RhPd]_{37}$	9.30	9.60	[41]
	$(ScZrNb)_{65}[RhPd]_{35}$	9.70	10.70	[41]
	$(ScZrNbTa)_{67}[RhPd]_{33}$	4.20	2.10	[41]
	$(ScZrNbTa)_{684}[RhPd]_{316}$	6.40	8.80	[41]
	$(ScZrNbTa)_{685}[RhPd]_{315}$	7.0		[42]
α-Mn	$(ZrNb)_{10}[MoReRu]_{90}$	5.30	7.86	[43]
	$(HfTaWIr)_{20}[Re]_{80}$	5.90	—	[43]
	$(HfTaWIr)_{40}[Re]_{60}$	4.00	4.64	[43]
	$(HfTaWPt)_{20}[Re]_{80}$	6.30	—	[43]
	$(HfTaWPt)_{40}[Re]_{60}$	4.40	5.90	[43]
	$(HfTaWPt)_{50}[Re]_{50}$	2.40	—	[43]
	$Nb_{25}Mo_5Re_{35}Ru_{25}Rh_{10}$	4.66	7.50	[44]
	$Nb_{25}Mo_{10}Re_{35}Ru_{20}Rh_{10}$	5.10	8.30	[44]
	$Nb_{25}Mo_{15}Re_{35}Ru_{15}Rh_{10}$	5.10	7.90	[44]
	$Nb_5Mo_{35}Re_{15}Ru_{35}Rh_{10}$	7.54	8.90	[44]
	$Nb_5Mo_{30}Re_{20}Ru_{35}Rh_{10}$	6.69	7.50	[44]
	$Nb_5Mo_{25}Re_{25}Ru_{35}Rh_{10}$	6.51	7.50	[44]
	$Nb_5Mo_{20}Re_{30}Ru_{35}Rh_{10}$	5.46	6.10	[44]
hcp	$Re_{56}Nb_{11}Ti_{11}Zr_{11}Hf_{11}$	4.40	3.60	[45]
	$(MoReRuRh)_{95}Ti_5$	3.60	—	[46]
	$(MoReRuRh)_{90}Ti_{10}$	4.70	—	[46]
	$Nb_{10}Mo_{35}Ru_{35}Rh_{10}Pd_{10}$	5.58	6.90	[47]
	$Nb_{15}Mo_{32.5}Ru_{32.5}Rh_{10}Pd_{10}$	6.19	8.10	[47]
	$Nb_{20}Mo_{30}Ru_{30}Rh_{10}Pd_{10}$	6.10	8.30	[47]
	$Mo_{30}Re_{29}Ru_{36}Pd_1Pt_4$	8.17	—	[49]
	$Mo_{27}Re_{25}Ru_{34}Pd_5Pt_9$	4.91	—	[49]
	$Mo_{24}Re_{20}Ru_{30}Pd_{12}Pt_{14}$	2.22	—	[49]
	$Mo_{22}Re_{22}Ru_{28}Pd_{10}Pt_{18}$	1.64	—	[49]
NiAs	$(RuRhPdIr)_{80}Pt_{20}Sb$ (SSR)	2.15	—	[48]
fcc	$Mo_{28}Re_{28}Ru_{35}Pd_4Pt_5C_{30}$	2.46	—	[49]
	$Mo_{29}Re_{27}Ru_{36}Pd_4Pt_4C_{44}$	2.72	—	[49]
	$Mo_{26}Re_{25}Ru_{33}Pd_8Pt_9C_{27}$	2.28	—	[49]

Structure	Nominal composition	T_c (K)	$B_{c2}(0)$ (T)	Reference
	$Mo_{27}Re_{25}Ru_{34}Pd_6Pt_9C_{40}$	2.30	—	[49]
	$Mo_{26}Re_{22}Ru_{31}Pd_9Pt_{13}C_{22}$	1.86	—	[49]
	$Mo_{23}Re_{23}Ru_{30}Pd_{11}Pt_{14}C_{31}$	1.91	—	[49]
	$Mo_{21}Re_{21}Ru_{27}Pd_{13}Pt_{18}C_{25}$	1.87	—	[49]
	$Mo_{21}Re_{22}Ru_{27}Pd_{14}Pt_{17}C_{30}$	1.75	—	[49]
σ-type	$Ta_5Mo_{35}W_5Re_{35}Ru_{20}$	6.29	—	[8]
	$Ta_5Mo_{30}W_{10}Re_{35}Ru_{20}$	6.20	—	[8]
	$Ta_5Mo_{25}W_{15}Re_{35}Ru_{20}$	6.10	—	[8]
	$Ta_5Mo_{20}W_{20}Re_{35}Ru_{20}$	5.69	—	[8]
	$Ta_5Mo_{15}W_{25}Re_{35}Ru_{20}$	5.46	—	[8]
	$Ta_5Mo_{10}W_{30}Re_{35}Ru_{20}$	5.45	—	[8]
	$Ta_5Mo_5W_{35}Re_{35}Ru_{20}$	4.78	—	[8]
	$Ta_7Mo_{33}W_5Re_{35}Ru_{20}$	6.13	—	[8]
	$Ta_9Mo_{31}W_5Re_{35}Ru_{20}$	5.70	—	[8]
	$Ta_{11}Mo_{29}W_5Re_{35}Ru_{20}$	5.32	—	[8]
	$Ta_{13}Mo_{27}W_5Re_{35}Ru_{20}$	5.27	—	[8]
	$Ta_{10}Mo_{30}Cr_5Re_{35}Ru_{20}$	4.79	6.10	[50]
	$Ta_{10}Mo_{25}Cr_{10}Re_{35}Ru_{20}$	4.41	5.80	[50]
	$Ta_{10}Mo_{22}Cr_{13}Re_{35}Ru_{20}$	3.98	4.90	[50]
	$Ta_{10}Mo_5W_{30}Re_{35}Ru_{20}$	4.87	6.70	[51]
	$Ta_{10}Mo_5W_{30}Re_{35}Ru_{20}C_2$	4.80	—	[51]
β-Mn	$Ta_{10}Mo_5W_{30}Re_{35}Ru_{20}C_{16}$	5.36	—	[51]
	$Ta_{10}Mo_5W_{30}Re_{35}Ru_{20}C_{18}$	5.32	—	[51]
	$Ta_{10}Mo_5W_{30}Re_{35}Ru_{20}C_{20}$	5.34	9.30	[51]
	$Cr_5Mo_{35}W_{12}Re_{35}Ru_{13}C_{20}$	5.49	9.70	[52]
	$Cr_8Mo_{32}W_{12}Re_{35}Ru_{13}C_{20}$	4.73	8.40	[52]
	$Cr_{11}Mo_{29}W_{12}Re_{35}Ru_{13}C_{20}$	3.83	7.30	[52]
	$Cr_{14}Mo_{26}W_{12}Re_{35}Ru_{13}C_{20}$	3.35	5.30	[52]
$CuAl_2$	$Mo_{11}W_{11}V_{11}Re_{34}B_{33}$	4.00	7.30	[53]

Table 4.
Selected properties of non-bcc structure-type HEA superconductors. SSR solid-state reaction; all other samples prepared by arc melting.

3. Superconducting high-entropy compounds

3.1 Intermetallics

Other high-entropy superconductors with different compositions present alternative structures to the previous ones, **Table 5**. For example, $(V_{0.5}Nb_{0.5})_{3-x}Mo_xAl_{0.5}Ga_{0.5}$ $(0.2 < x \leq 1.4)$ is polymorphic with bcc and A15 structure (similar to α- and β-W

Structure	Nominal composition	T_c (K)	$B_{c2}(0)$ (T)	Reference
A15	$(V_{0.5}Nb_{0.5})_{2.8}Mo_{0.2}Al_{0.5}Ga_{0.5}$	10.20	20.10	[7]
	$(V_{0.5}Nb_{0.5})_{2.6}Mo_{0.4}Al_{0.5}Ga_{0.5}$	9.20	17.70	[7]
	$(V_{0.5}Nb_{0.5})_{2.4}Mo_{0.6}Al_{0.5}Ga_{0.5}$	8.90	17.00	[7]
	$(V_{0.5}Nb_{0.5})_{2}Mo_{1}Al_{0.5}Ga_{0.5}$	6.10	9.90	[7]
	$(V_{0.5}Nb_{0.5})_{1.8}Mo_{1.2}Al_{0.5}Ga_{0.5}$	4.80	7.60	[7]
	$(V_{0.5}Nb_{0.5})_{1.6}Mo_{1.4}Al_{0.5}Ga_{0.5}$	3.20	4.80	[7]
	$V_5Nb_{35}Mo_{35}Ir_{10}Pt_{15}$	5.18	6.40	[55]
	$V_{15}Nb_{30}Mo_{30}Ir_{10}Pt_{15}$	4.49	5.70	[55]
	$V_{25}Nb_{25}Mo_{25}Ir_{10}Pt_{15}$	3.61	4.40	[55]
	$Nb_3Al_{0.2}Sn_{0.2}Ge_{0.2}Ga_{0.2}Si_{0.2}$	9.00	10.40	[56]
	$Nb_3Al_{0.3}Sn_{0.3}Ge_{0.2}Ga_{0.1}Si_{0.1}$	11.00	13.30	[56]
	$V_3Al_{0.07}Si_{0.30}Ga_{0.08}Ge_{0.30}Sn_{0.25}$	6.30	8.8	[57]
$CuAl_2$	$Co_{0.2}Ni_{0.1}Cu_{0.1}Rh_{0.3}Ir_{0.3}Zr_2$	8.0	12	[58]
	$Fe_{0.09}Co_{0.19}Ni_{0.11}Rh_{0.27}Ir_{0.33}Zr_2$	7.8	8.5	[54]
	$Fe_{0.11}Co_{0.30}Ni_{0.20}Rh_{0.07}Ir_{0.32}Zr_2$	6.7	6.6	[54]
	$Fe_{0.19}Co_{0.19}Ni_{0.20}Rh_{0.21}Ir_{0.21}Zr_2$	5.4	5.5	[54]
	$Fe_{0.29}Co_{0.19}Ni_{0.30}Rh_{0.09}Ir_{0.12}Zr_2$	4.8	4.8	[54]
	$Fe_{0.09}Co_{0.20}Ni_{0.20}Cu_{0.17}Rh_{0.19}Ir_{0.15}Zr_2$	5.7		[59]

Table 5.
Selected properties of A15 and $TrZr_2$ high-entropy superconductors. All samples have been prepared by arc melting. The compositions of Ref. [54] were rounded to two decimals for simplicity.

actually) [7]. Furthermore, heating also induces a polymorphic transformation from original bcc structure to the A15 one, as shown for $x = 0.2$. The resulting compound with this A15 structure shows bulk superconductivity with $T_c = 10.2$ K and an estimated $B_{c2}(0) = 20.1$ T [52]. Here, mixing on both distinct lattice sites occurs, and a true alloy resembling β-W is forming. A similar case is the system $V_{5+2x}Nb_{35-x}Mo_{35-x}Ir_{10}Pt_{15}$ $(0 \leq x \leq 10)$ with maximum $T_c = 5.18$ K and $B_{c2}(0) = 6.4$ T for $x = 0$ [55].

The A15 structure type can be also found in high-entropy compounds like $Nb_3Al_{0.2}Sn_{0.2}Ge_{0.2}Ga_{0.2}Si_{0.2}$ with a $T_c = 9.0$ K and $B_{c2}(0) = 10.4$ T [56], $Nb_3Al_{0.3}Sn_{0.3}Ge_{0.2}Ga_{0.1}Si_{0.1}$ with a $T_c = 11.0$ K and $B_{c2}(0) = 13.3$ T [56], and $V_3Al_{0.07}Si_{0.30}Ga_{0.08}Ge_{0.30}Sn_{0.25}$ with a $T_c = 6.3$ K and $B_{c2}(0) = 8.8$ T [57]. In all these cases, mixing occurs primarily on just one of the lattice sites (disregarding natural disorder), and a compound resembling Cr_3Si (or Nb_3Sn, for that matter) is formed. The V system is also interesting because it introduced the concept of compositionally complex alloys (CCAs) as a new category of multi-principal-element materials closely related to high-entropy alloys, but containing more than two phases in an alloy sample. In the particular case of $V_3Al_{0.07}Si_{0.30}Ga_{0.08}Ge_{0.30}Sn_{0.25}$, it crystallizes into five high-entropy A15 phases with different compositions. This concept of CCAs is thought to be a practical way to improve the properties of the superconducting HEAs.

Further high-entropy intermetallics are the so-called $TrZr_2$ compounds (Tr (mix of) transition metal) with $CuAl_2$ crystal structure. They have been discovered by

Mizuguchi et al. in $Co_{0.2}Ni_{0.1}Cu_{0.1}Rh_{0.3}Ir_{0.3}Zr_2$ and can be regarded as mixing of single-*Tr* zirconides, of which many are superconductors themselves [58]. This high-entropy compound shows a T_c of 8.0 K and a $B_{c2}(0)$ of 12 T. Shortly after, superconductivity was also found in (Fe,Co,Ni,Rh,Ir)Zr_2 samples of different stoichiometries [54]. The same group found an unusual broadening of the specific heat jump at T_c for this system (developing with the number of *Tr* elements), which they attributed to an inhomogeneity of the Cooper pair formation and hence the superconducting gap [59]. Pugliese et al. [60] investigated possible local inhomogeneities in these systems by extended X-ray absorption fine structure (EXAFS) measurements and found a correlation between T_c and the Zr 4*d*-*Tr* n*d* hybridization. The Zr-Zr bonds seem to get stiffer with increasing mixing entropy, which suggests nanoscale texturing with possible local order.

3.2 Non-oxide ceramics

The concept of high-entropy alloys primarily focused on metal alloys, however, was quickly extended to other types of materials, **Table 6**. This approach, centered on maximizing the configurational entropy to stabilize equimolar or near-equimolar mixtures, was employed in a mixture of oxides to synthesize the first high-entropy oxide (HEO) in 2015 [71], giving rise to a new category known as high-entropy ceramics (HECs). HECs encompass all high-entropy materials with ceramic properties. Similar to HEAs, HECs consist of multicomponent elements in a single phase,

Structure	Nominal composition	T_c (K)	$B_{c2}(0)$ (T)	Reference
NaCl	(TiZrNbHfTa)C	2.35	0.51	[62]
	$(MoNbTaVW)C_{0.9}$	3.4	3.37	[63]
	(AgInSnPbBi)Te	2.6	2.8	[64]
	$(\{AgSnPbBi\}_{(1-x)/4}In_x)Te$	<2.8	—	[65]
1 T-$NiTe_2$	$(Co,Au)_{0.2}(Rh,Ir,Pd,Pt)_{0.8}Te_2$	4.5	—	[66]
	$Co_{0.03}Au_{0.06}Rh_{0.23}Ir_{0.24}Pd_{0.16}Pt_{0.28}Te_2$	2.5	—	[66]
W_5Si_3	$(Nb_{0.1}Mo_{0.3}W_{0.3}Re_{0.2}Ru_{0.1})_5Si_3$	3.30	5.00	[67]
	$(Nb_{0.2}Mo_{0.3}W_{0.3}Re_{0.1}Ru_{0.1})_5Si_3$	3.20	5.10	[67]
LaOBiS2	$La_{0.1}Ce_{0.1}Pr_{0.2}Nd_{0.3}Sm_{0.3}O_{0.5}F_{0.5}BiS_2$	4.9		[68]
	$La_{0.2}Ce_{0.2}Pr_{0.2}Nd_{0.2}Sm_{0.2}O_{0.5}F_{0.5}BiS_2$	3.97		[69]
	$La_{0.3}Ce_{0.3}Pr_{0.2}Nd_{0.1}Sm_{0.1}OBiS_2$	3.4	0.65/11	[61]
	$La_{0.1}Ce_{0.3}Pr_{0.3}Nd_{0.2}Sm_{0.1}OBiS_2$	4.3	0.69/15	[61]
	$La_{0.2}Ce_{0.2}Pr_{0.2}Nd_{0.2}Sm_{0.2}OBiS_2$	3.3	0.24/4.9	[61]
	$La_{0.1}Ce_{0.3}Pr_{0.1}Nd_{0.2}Sm_{0.3}OBiS_2$	4.6	0.42/16	[61]
	$La_{0.1}Ce_{0.3}Pr_{0.3}Nd_{0.1}Sm_{0.2}OBiS_2$	3.7		[70]
	$La_{0.2}Ce_{0.3}Pr_{0.3}Sm_{0.1}Gd_{0.1}OBiS_2$	3.0		[70]
	$La_{0.2}Ce_{0.3}Pr_{0.2}Nd_{0.1}Sm_{0.1}Gd_{0.1}OBiS_2$	2.9		[70]

Table 6.
Selected properties of high-entropy superconducting ceramics. For Ref. [61], $B_{c2}(0)$ *is given for both major field directions, B||c and B⊥c.*

where their substantial configurational entropy plays a crucial role in their formation. In contrast to metallic HEAs, HECs typically exhibit semiconductor or insulator characteristics, making them potentially useful as functional materials. These systems have proven to be versatile in various technologies, including thermal barrier coatings, thermoelectrics, catalysts, and batteries, as well as wear-resistant and corrosion-resistant coatings [72–74]. However, in this chapter, the focus will be on those materials that show superconducting properties.

High-entropy carbide ceramics (HECCs) are a subgroup of materials belonging to the more general one of the HEC. The high-entropy carbides are coming from the binary transition metal carbides (*TM*Cs), composed of carbon and a transition metal, that are widely used as ultra-high-temperature ceramics in structural applications. They are made by combining different *TM*Cs to synthesize a single-phase multi-principal elemental HECC [75]. Among them, (TiZrNbHfTa)C was the first one to show superconducting properties [62]. This HECC, prepared by SPS, crystallizes in NaCl structure and shows bulk superconductivity with T_c = 2.35 K, $B_{c1}(0) \sim 26.1$ mT, and $B_{c2}(0) \sim 0.51$ T. Those superconducting properties are robust and were kept almost invariable under pressure up to 80 GPa [74]. Very recently, Shu et al. discovered the coexistence of ferromagnetism and superconductivity in $(MoNbTaVW)C_{0.9}$ and (TaTiNbZr)C [63].

Some superconducting telluride high-entropy compounds crystallize in the NaCl structure too (**Figure 2b**). Superconductivity was first seen in $AgInSnPbBiTe_5$ by Mizuguchi with a T_c of ~2.6 K and an extrapolated $B_{c2}(0)$ of ~2.8 T [64]. Upon applying pressure up to 35 GPa, the crystal structure changes via a Pnma structure (at around 12 GPa) to the CsCl type with increased T_c values of ~5.3 K [76]. The stoichiometry on the *M* site was varied shortly after by adjusting the In content as $(AgSnPbBi)_{(1-x)/4}In_xTe$ with x = 0, 0.1, 0.2, 0.3, and 0.4, where the sample without In is not superconducting, and T_c increases slightly with x [65]. A special case of superconducting high-entropy tellurides is $(Co,Au)_{0.2}(Rh,Ir,Pd,Pt)_{0.8}Te_2$ and $Co_{0.03}Au_{0.06}Rh_{0.23}Ir_{0.24}Pd_{0.16}Pt_{0.28}Te_2$ as layered compounds with 1 T-$NiTe_2$ structure. They are called high-entropy van der Waals materials (HEX) and show superconductivity up to 4.5 K [66].

The case of the high-entropy silicides (HESs) is quite similar to the carbides. They are made by combining Si with other elements, mainly metals. The binary transition-metal silicides M_5Si_3 were studied extensively for high-temperature structural applications due to their high melting points. Again, combining these binary silicides, different HESs were synthesized. However, the first superconducting high-entropy silicides appeared only very recently [67]. $(Nb_{0.1}Mo_{0.3}W_{0.3}Re_{0.2}Ru_{0.1})_5Si_3$ and $(Nb_{0.2}Mo_{0.3}W_{0.3}Re_{0.1}Ru_{0.1})_5Si_3$ were prepared by arc melting and are weakly coupled bulk superconductors. They both crystallize in the W_5Si_3 structure and have T_c values of 3.3 K and 3.2 K, and $B_{c2}(0)$ values of 5.0 T and 5.1 T, respectively [65].

Superconducting high-entropy sulfides are derived from the layered superconductors $RE(O,F)BiS_2$ by mixing the rare earth element (*RE*) (**Figure 2c**) [77]. The first reported system was $(La,Pr,Ce,Nd,Sm)O_{0.5}F_{0.5}BiS_2$, where T_c depends linearly on lattice parameter a, and $La_{0.1}Ce_{0.1}Pr_{0.2}Nd_{0.3}Sm_{0.3}O_{0.5}F_{0.5}BiS_2$ showed the highest value of ~4.9 K [68]. The same group investigated shortly after the dependence of structural and superconducting properties on the mixing entropy by compounds with one to five *RE* elements, where T_c and superconducting volume fraction increase linearly with mixing entropy [69]. Shortly after, superconductivity was also confirmed in F-free *HE RE*$OBiS_2$ single crystals [61, 70], for which the anisotropy of the upper critical field could be determined.

3.3 Oxides

The first high-entropy oxides (HEOs) were reported by Rost et al. in 2015 [71]. In this work, the authors demonstrated that the entropy drives a reversible solid-state transformation between a multiphase and a single-phase state and was able to synthesize $(Mg_{0.2}Co_{0.2}Ni_{0.2}Cu_{0.2}Zn_{0.2})O$, the first HEO [61]. However, the term HEO was only introduced 1 year later by Bérardan et al., making the analogy to HEAs, to classify these multicationic, equiatomic oxide systems [78]. Over the years, the HEOs have gained significant interest due to their unique structural characteristics and related possibilities for tailoring functional properties [79–82]. One particularity of the HEOs, in contrast to HEAs, is that here the elemental mixing is restricted to one or only some of the atomic positions, similar to the metallic and ceramic compounds mentioned above. Superconductivity in HEOs could only be found in compounds of the $REBa_2Cu_3O_{7-\delta}$ (*RE*BCO) family so far, probably the best-known and studied superconducting oxides. A recent search, for example, for superconducting high-entropy Ruddlesden-Popper cuprates has not been successful so far [83].

The concept of HEO was applied for the first time to *RE*BCO compounds in 2020 by Shukunami et al. when the first high-entropy *RE*BCO (HE *RE*BCO) compound was reported [10]. The *RE*BCO compounds exhibit a triple perovskite structure, whose center is the *RE* atom. This *RE* site is occupied with a mixture of rare earth elements, so that is where entropy is introduced (**Figure 2d**).

To the best of our knowledge, although the number has been increasing year by year since 2020, only 10 articles have been reported on HE *RE*BCO to date. The different HE *RE*BCO compounds presented in those articles can be classified as bulks or films according to the synthesis method followed (**Table** 7). Within the group of the films, one can further distinguish between films by pulsed laser deposition (PLD) and by chemical solution deposition (CSD).

Shukunami et al. manufactured polycrystalline *RE*BCO bulk samples via solid-state reaction, combining up to six different rare earths in 2020 [10]. So far, all bulk HE *RE*BCO samples have been prepared by SSR. In particular, they presented $Y_{0.28}Nd_{0.16}Sm_{0.18}Eu_{0.18}Gd_{0.20}Ba_2Cu_3O_{7-\delta}$ and $Y_{0.18}La_{0.24}Nd_{0.14}Sm_{0.14}Eu_{0.15}Gd_{0.15}Ba_2Cu_3O_{7-\delta}$ as the first two superconducting HE *RE*BCO compounds with T_c = 93 K in both cases. The J_c values of these compounds were calculated from magnetic measurements at 2 K and 1 T. The values were 11.7 kA/cm^2 for $Y_{0.28}Nd_{0.16}Sm_{0.18}Eu_{0.18}Gd_{0.20}Ba_2Cu_3O_{7-\delta}$ (~20 kA/cm^2 at self-field) and 8.1 kA/cm^2 for $Y_{0.18}La_{0.24}Nd_{0.14}Sm_{0.14}Eu_{0.15}Gd_{0.15}Ba_2Cu_3O_{7-\delta}$. The authors also identify a clear correlation between T_c and the orthorhombicity parameter (*OP*), which is defined as $2|a - b|/(a+b)$ with lattice constants a and b. For $OP > 0.012$, the T_c of the samples is almost constant, but for $OP < 0.012$, T_c starts to decay. For $OP < 0.007$, superconductivity was not observed anymore. Interestingly and accordingly, two further HE *RE*BCO compounds that they prepared were not superconducting, namely, $Y_{0.23}La_{0.25}Nd_{0.15}Sm_{0.19}Eu_{0.18}Ba_2Cu_3O_{7-\delta}$ and the $Y_{0.23}La_{0.23}Nd_{0.17}Sm_{0.18}Gd_{0.19}Ba_2Cu_3O_{7-\delta}$. These results indicate the importance of wisely selecting the *RE* elements to include in the HE *RE*BCO to maximize the superconducting properties. One year later, Wang et al. [84] prepared four new HE *RE*BCO compounds with a mixture of five elements in each case: $Y_{0.2}La_{0.2}Sm_{0.2}Dy_{0.2}Er_{0.2}Ba_2Cu_3O_{7-\delta}$, $Y_{0.2}Ho_{0.2}Er_{0.2}Tm_{0.2}Yb_{0.2}Ba_2Cu_3O_{7-\delta}$, $Y_{0.5}La_{0.125}Sm_{0.125}Dy_{0.125}Er_{0.125}Ba_2Cu_3O_{7-\delta}$, and $Y_{0.5}Sm_{0.125}Eu_{0.125}Gd_{0.125}Dy_{0.125}Ba_2Cu_3O_{7-\delta}$. Magnetic measurements showed that all these samples exhibit a very similar T_c of ~92 K. The self-field J_c values of

Nominal composition on *RE* site	T_c (K)	J_c (MA/cm^2)	Reference
$Y_{0.28}Nd_{0.16}Sm_{0.18}Eu_{0.18}Gd_{0.20}$	93.0	0.02 (2 K, sf) 0.0117 (2 K, 1 T)	[10]
$Y_{0.18}La_{0.24}Nd_{0.14}Sm_{0.14}Eu_{0.15}Gd_{0.15}$	93.0	0.081 (2 K, 1 T)	[10]
$Y_{0.2}La_{0.2}Sm_{0.2}Dy_{0.2}Er_{0.2}$	92.0	—	[84]
$Y_{0.2}Ho_{0.2}Er_{0.2}Tm_{0.2}Yb_{0.2}$	92.0	—	[84]
$Y_{0.5}La_{0.125}Sm_{0.125}Dy_{0.125}Er_{0.125}$	92.0	0.02 (2 K, sf) 0.002 (77 K, sf)	[84]
$Y_{0.5}Sm_{0.125}Eu_{0.125}Gd_{0.125}Dy_{0.125}$	92.0	—	[84]
$Y_{0.2}La_{0.2}Sm_{0.2}Eu_{0.2}Gd_{0.2}$	93.4	0.0109 (J_c^{Global} at 2 K, 1 T) 6.5 (J_c^{local} at 2 K, sf)	[85]
$Y_{0.2}Nd_{0.2}Sm_{0.2}Eu_{0.2}Gd_{0.2}$	93.4	0.0117 (2 K, 1 T)	[85]
$Y_{0.2}Sm_{0.2}Eu_{0.2}Dy_{0.2}Ho_{0.2}$	93.0	0.0346 (2 K, 1 T)	[85]
$Y_{1/6}La_{1/6}Nd_{1/6}Sm_{1/6}Eu_{1/6}Gd_{1/6}$	93.1	0.0081 (2 K, 1 T)	[85]
$Y_{1/6}Sm_{1/6}Eu_{1/6}Dy_{1/6}Ho_{1/6}Yb_{1/6}$	92.0	0.0499 (2 K, 1 T)	[85]
$Y_{1/7}Sm_{1/7}Eu_{1/7}Dy_{1/7}Ho_{1/7}Yb_{1/7}Lu_{1/7}$	91.4	0.0535 (J_c^{Global} at 2 K, 1 T) 5.5 (J_c^{local} at 2 K, sf)	[85]
$Dy_{0.16}Ho_{0.17}Er_{0.20}Tm_{0.22}Yb_{0.25}$	88.8	—	[86]
$Gd_{0.23}Dy_{0.17}Ho_{0.15}Er_{0.15}Tm_{0.19}Yb_{0.20}$	89.8	—	[86]
$Gd_{0.20}Dy_{0.13}Ho_{0.10}Er_{0.15}Tm_{0.09}Yb_{0.20}Lu_{0.15}$	89.4	—	[86]
$Y_{0.12}Sm_{0.08}Eu_{0.26}Dy_{0.18}Ho_{0.36}$ (PLD film)	90.5	6.5 (2 K, 1 T) 2.3 (4.2 K, 7 T)	[87]
$Y_{0.7}Gd_{0.2}Dy_{0.2}Sm_{0.2}Eu_{0.2}$ (CSD film)	93.0	10 (30 K, sf) 4 (65 K, sf) 2 (77 K, sf)	[88]
$Gd_{0.2}Sm_{0.2}Nd_{0.2}Eu_{0.2}Y_{0.2}$ (CSD film)	90.4	2.4 (77 K, sf)	[89]
$Gd_{0.2}Dy_{0.2}Y_{0.2}Ho_{0.2}Er_{0.2}$ (+12% BHO, CSD)	91.9	3.5 (77 K, sf)	[90]

Table 7.
Critical temperatures and selected J_c *data of high-entropy* REBCO *phases. The samples, except otherwise stated, are prepared by solid-state reaction. sf, self-field or remanent state.*

$Y_{0.5}La_{0.125}Sm_{0.125}Dy_{0.125}Er_{0.125}Ba_2Cu_3O_{7-\delta}$ were 20 kA/cm^2 at 1.8 K and 2 kA/cm^2 at 77 K. These samples show smaller irreversible fields than traditional YBCO, indicating again that the selection of the *RE* elements has a huge influence on the properties of these compounds. In 2022, Yamashita et al. of the same group that reported the first HE *RE*BCO went a step further [85]. They prepared six new compounds by solid-state reaction including one with seven different *RE*: $Y_{0.2}La_{0.2}Sm_{0.2}Eu_{0.2}Gd_{0.2}Ba_2Cu_3O_{7-\delta}$, $Y_{0.2}Nd_{0.2}Sm_{0.2}Eu_{0.2}Gd_{0.2}Ba_2Cu_3O_{7-\delta}$, $Y_{0.2}Sm_{0.2}Eu_{0.2}Dy_{0.2}Ho_{0.2}Ba_2Cu_3O_{7-\delta}$, $Y_{1/6}La_{1/6}Nd_{1/6}Sm_{1/6}Eu_{1/6}Gd_{1/6}Ba_2Cu_3O_{7-\delta}$, $Y_{1/6}Sm_{1/6}Eu_{1/6}Dy_{1/6}Ho_{1/6}Yb_{1/6}Ba_2Cu_3O_{7-\delta}$, and $Y_{1/7}Sm_{1/7}Eu_{1/7}Dy_{1/7}Ho_{1/7}Yb_{1/7}Lu_{1/7}Ba_2Cu_3O_{7-\delta}$. T_c values of these compounds are between 93.4 K and 91.4 K, slightly decreasing with *RE* complexity. Similar to their first work on T_c, they correlated the *OP* here with J_c. Again, there is an obvious correlation between both parameters, being the largest J_c = 53.5 kA/cm^2 (2 K, 1 T), for the largest *OP* that corresponds to the compound with seven *RE*, even though T_c was the smallest. This value of J_c, measured magnetically on cube-shaped samples, corresponds to J_c^{Global}. However, this J_c is affected by the typical weak links from grain to grain found in the *RE*BCO compounds due to their short coherence lengths. Therefore, the authors estimate J_c within the grains (J_c^{local}) in fine powders of the same composition as the samples by magnetic measurements. At 2 K, J_c^{local} of ~6.5 MA/cm^2 was found for $Y_{0.2}Sm_{0.2}Eu_{0.2}Dy_{0.2}Ho_{0.2}Ba_2Cu_3O_{7-\delta}$ and ~ 5.5 MA/cm^2 for $Y_{1/7}Sm_{1/7}Eu_{1/7}Dy_{1/7}Ho_{1/7}Yb_{1/7}Lu_{1/7}Ba_2Cu_3O_{7-\delta}$. At 20 K, the values for the different compounds are very similar, ~2 MA/cm^2 independent of the number of *RE* [84]. Also, in 2023, the use of solid-state reaction synthesis has been continued. Suzuki et al. [86] followed this method to prepare HE *RE*BCO single-crystal whiskers with dimensions 0.5–1.0 mm in length, 10 μm in thickness, and 10–50 μm in width. The analytical HE *RE*BCO compositions were $Dy_{0.16}Ho_{0.17}Er_{0.20}Tm_{0.22}Yb_{0.25}Ba_2Cu_3O_{7-\delta}$, $Gd_{0.23}Dy_{0.17}Ho_{0.15}Er_{0.15}Tm_{0.19}Yb_{0.20}Ba_2Cu_3O_{7-\delta}$, and $Gd_{0.20}Dy_{0.13}Ho_{0.10}Er_{0.15}Tm_{0.09}Yb_{0.20}Lu_{0.15}Ba_2Cu_3O_{7-\delta}$. The T_c zero values of these compounds were around 89 K, which indicates that T_c of HE *RE*BCO compounds with a similar average *RE* ionic radius does not depend on the mixture entropy at the *RE* site. Therefore, the advantages of high-entropy alloys may be kept for a high degree of entropy without reducing the superconducting properties of the *RE*BCO compounds. Finally, Pryanichnikov et al. [91] synthesized equimolar $Y_{0.2}Nd_{0.2}Eu_{0.2}Sm_{0.2}Ho_{0.2}Ba_2Cu_3O_y$ bulk samples and studied their superconducting properties. The purpose of this work was to change the oxygen content by annealing in appropriate conditions and to study the crystal structure features and superconducting properties. It was concluded that *OP*, T_c, and J_c of $Y_{0.2}Nd_{0.2}Eu_{0.2}Sm_{0.2}Ho_{0.2}Ba_2Cu_3O_y$ increase with oxygen content y, very similar to their relatives of "low-entropy" *RE*BCO compounds.

Within the group of HE *RE*BCO compounds prepared as films, the first work published was the one of Yamashita et al., who prepared $Y_{0.12}Sm_{0.08}Eu_{0.26}Dy_{0.18}Ho_{0.36}Ba_2Cu_3O_{7-\delta}$ films by PLD [87]. This sample containing five *RE*s was compared with other low- and medium-entropy compounds. All samples grew with *c*-axis-preferential orientation. In addition, the *c*-axis of the samples increases with the average diameter of the *RE*s mixture and so does the width of the (005) peak, indicating, as expected, that there is an increase in the disorder of the crystal lattice. The T_c of the $Y_{0.12}Sm_{0.08}Eu_{0.26}Dy_{0.18}Ho_{0.36}Ba_2Cu_3O_{7-\delta}$ film was 90.5 K, the same as for the medium-entropy films with three and four *RE*s. It shows J_c = 2.3 MA/cm^2 at 4.2 K and 7 T, and that stays above 1.0 MA/cm^2 up to 20 K and 7 T. These values are very similar to the medium-entropy film with four *RE* and to the standard YBCO film in the high-field region. Few months later, Chen et al. reported the first HE *RE*BCO film deposited by fluorine-free CSD [88], namely, an 800-nm thick nominal $Y_{0.7}Gd_{0.2}Dy_{0.2}Sm_{0.2}Eu_{0.2}Ba_2Cu_3O_{7-\delta}$ film on technical substrates. This compound presents the particularity that it was nominally off stoichiometry on the *RE* molarity because the authors argued this type of off-stoichiometric composition ratio provided better superconducting properties than standard 1:2:3, possibly mostly due to the formation of RE_2O_3 nanoparticles, which serve as pinning centers. The structural and superconducting properties of this HE *RE*BCO were compared with two control samples prepared the same way, $Y_1Dy_{0.5}Ba_2Cu_3O_{7-\delta}$ (also off stoichiometry on the *RE* molarity) and YBCO. They showed that high-quality *c*-axis-oriented $Y_{0.7}Gd_{0.2}Dy_{0.2}Sm_{0.2}Eu_{0.2}Ba_2Cu_3O_{7-\delta}$ films with T_c = 93 K can be obtained. The in-field J_c values of this HE *RE*BCO film are significantly improved, especially at 77 K or below 4 T. Self-field J_c values of 2, 4, and 10 MA/cm^2 were obtained at 77, 65, and 30 K, respectively. The maximum pinning force density of the HE *RE*BCO film at 77 K was 1.7 times higher than that of the standard YBCO film. However, it is difficult to confirm if this enhancement of J_c originates from the effect of the HE *RE* mixing or from the contribution of the RE_2O_3 pinning centers that were also found in the films. Fluorine-free solutions were also used by Masuda et al. to deposit 300-nm $Gd_{0.2}Sm_{0.2}Nd_{0.2}Eu_{0.2}Y_{0.2}Ba_2Cu_3O_{7-\delta}$ films on $LaAlO_3$ substrates [89]. The films show a T_c = 90.4 K and a self-field J_c = 2.4 MA/cm^2 at 77 K, both exceeding a standard GdBCO film of the same study. Cayado et al. raised the level of complexity further by introducing perovskite nanoparticles into the HE *RE*BCO film [90]. They prepared $Gd_{0.2}Dy_{0.2}Y_{0.2}Ho_{0.2}Er_{0.2}Ba_2Cu_3O_{7-\delta}$ nanocomposite films with 12 mol% $BaHfO_3$ (BHO) nanoparticles on $SrTiO_3$ substrates by CSD using full-TFA solutions. Both T_c and self-field J_c at 77 K of such a film, 91.9 K and 3.5 MA/cm^2, respectively, are comparable with the ones of low- and medium-entropy *RE*BCO films prepared in the same way. By an exhaustive TEM study, a homogeneous distribution of the BHO nanoparticles as well as a homogeneous distribution of the *RE*s at the atomic level instead of *RE* clustering, were observed, proving a true HE *RE*BCO. Grünewald et al. [92] of the same group continued these TEM analyses of such films, focusing on the mixing of *RE*s into the BHO nanoparticles. This is important as the possible incorporation of *RE*s in the nanoparticles could lead to off-stoichiometries in the *RE*BCO phase, causing a deviation from the expected growth process and reduction of the superconducting properties. The degree of *RE*s incorporation in BHO was found to depend on both the ionic radius and the content of the *RE*s in the *RE*BCO phase. The *RE*s incorporation increases with decreasing ionic radii and shows a stronger tendency for smaller ionic radii to intermix in medium- and high-entropy *RE*BCO than in the low-entropy ones.

4. Conclusion

In this comprehensive review, we have described the state of the art of superconducting high-entropy materials, focusing on fabrication methods and superconducting properties. The extension of high-entropy principles to the field of superconductivity has opened up new frontiers in the search for new superconductors and offers promising perspectives for innovative technological advancements. The potential applications of superconducting high-entropy materials cover a wide spectrum, from energy transmission to quantum computing. Their ability to maintain superconductivity under varying external conditions positions them as promising candidates for different applications. The discovery of new high-entropy superconductors has been continuous as the first one was found in 2014, while the interest in this topic is growing year by year. Due to their chemical similarities, transition metals and rare earth elements are predestined for mixing on certain lattice sites. The first superconducting HE alloys contained only low transition metals, which were later extended to higher transition metals and even main group metals. These alloys with simple bcc crystal structures were further developed toward more complex crystal structures and further to compounds. In two layered compound classes, *RE*OBiS$_2$ and *RE*BCO, high-entropy mixing on the *RE* site has been shown. Similar investigations might be interesting in the iron-based superconducting *Ln*FeAsO system, *Ln* lanthanoid, which is still missing. Combining transition metal mixing and rare earth mixing on appropriate lattice sites might be a further increase in complexity. There is certainly still plenty of room for new, exciting investigations in this field.

Author details

Pablo Cayado[1*] and Jens Hänisch[2]

1 Department of Quantum Matter Physics (DQMP), University of Geneva, Geneva, Switzerland

2 Karlsruhe Institute of Technology (KIT), Institute for Technical Physics (ITEP), Eggenstein-Leopoldshafen, Germany

*Address all correspondence to: pablo.cayado@unige.ch

References

[1] Yeh JW, Chen SK, Lin SJ, Gan JY, Chin TS, Shun TT, et al. Nanostructured high-entropy alloys with multiple principal elements: Novel alloy design concepts and outcomes. Advanced Engineering Materials. 2004;**6**(5):299-303

[2] Cantor B, Chang ITH, Knight P, Vincent AJB. Microstructural development in equiatomic multicomponent alloys. Materials Science and Engineering: A. 2004;**375–377**(1-2 Spec. issue):213-218

[3] Yuan Y, Wu Y, Luo H, Wang Z, Liang X, Yang Z, et al. Superconducting Ti15Zr15Nb35Ta35 high-entropy alloy with intermediate electron-phonon coupling. Frontiers in Materials. 2018;**5**

[4] Sun L, Cava RJ. High-entropy alloy superconductors: Status, opportunities, and challenges. Physical Review Materials. 2019;**3**(9):090301

[5] Koželj P, Vrtnik S, Jelen A, Jazbec S, Jagličić Z, Maiti S, et al. Discovery of a superconducting high-entropy alloy. Physical Review Letters. 2014;**113**(10):107001

[6] Matthias BT. Empirical relation between superconductivity and the number of valence electrons per atom. Physical Review. 1955;**97**:74

[7] Wu J, Liu B, Cui Y, Zhu Q, Xiao G, Wang H, et al. Polymorphism and superconductivity in the V-Nb-Mo-Al-Ga high-entropy alloys. Science China Materials. 2020;**63**(5):823-831

[8] Liu B, Wu J, Cui Y, Zhu Q, Xiao G, Wang H, et al. Formation and superconductivity of single-phase high-entropy alloys with a tetragonal structure. ACS Applied Electronic Materials. 2020;**2**(4):1130-1137

[9] Senkov ON, Scott JM, Senkova SV, Miracle DB, Woodward CF. Microstructure and room temperature properties of a high-entropy TaNbHfZrTi alloy. Journal of Alloys and Compounds. 2011;**509**(20):6043-6048

[10] Shukunami Y, Yamashita A, Goto Y, Mizuguchi Y. Synthesis of RE123 high-Tc superconductors with a high-entropy-alloy-type RE site. Physica C: Superconductivity and Its Applications. 2020;**572**:1353623

[11] Jasiewicz K, Wiendlocha B, Korbeń P, Kaprzyk S, Tobola J. Superconductivity of Ta34Nb33Hf8Zr14Ti11 high entropy alloy from first principles calculations. Physica Status Solidi (RRL)–Rapid Research Letters. 2016;**10**(5):415-419

[12] Vrtnik S, Koželj P, Meden A, Maiti S, Steurer W, Feuerbacher M, et al. Superconductivity in thermally annealed Ta-Nb-Hf-Zr-Ti high-entropy alloys. Journal of Alloys and Compounds. 2017;**695**:3530-3540

[13] Von Rohr F, Winiarski MJ, Tao J, Klimczuk T, Cava RJ. Effect of electron count and chemical complexity in the Ta-Nb-Hf-Zr-Ti high-entropy alloy superconductor. Proceedings of the National Academy of Sciences. 2016;**113**(46):E7144-E7150

[14] Guo J, Wang H, Von Rohr F, Wang Z, Cai S, Zhou Y, et al. Robust zero resistance in a superconducting high-entropy alloy at pressures up to 190 GPa. Proceedings of the National Academy of Sciences. 2017;**114**(50):13144-13147

[15] Wu KY, Chen SK, Wu JM. Superconducting in equal molar NbTaTiZr-based high-entropy alloys. Natural Science. 2018;**10**(03):110

[16] Kim G, Lee MH, Yun JH, Rawat P, Jung SG, Choi W, et al. Strongly correlated and strongly coupled s-wave superconductivity of the high entropy alloy Ta1/6Nb2/6Hf1/6Zr1/6Ti1/6 compound. Acta Materialia. 2020;**186**: 250-256

[17] Kim J, Jung SG, Han Y, Kim JH, Rhyee JS, Yeo S, et al. Thermal-driven gigantic enhancement in critical current density of high-entropy alloy superconductors. Journal of Materials Science and Technology. 2024;**189**:60-67

[18] Idczak R, Nowak W, Rusin B, Topolnicki R, Ossowski T, Babij M, et al. Enhanced superconducting critical parameters in a new high-entropy alloy Nb0.34Ti0.33Zr0.14Ta0.11Hf0.08. Materials. 2023;**16**(17):5814

[19] Jasiewicz K, Wiendlocha B, Górnicka K, Gofryk K, Gazda M, Klimczuk T, et al. Pressure effects on the electronic structure and superconductivity of (TaNb)0.67(HfZrTi)0.33 high entropy alloy. Physical Review B. 2019;**100**(18): 184503

[20] Horvat J, Soltanian S, Pan AV, Wang XL. Superconducting screening on different length scales in high-quality bulk MgB2 superconductor. Journal of Applied Physics. 2004;**96**(8):4342-4351

[21] Kim JH, Hidayati R, Jung SG, Salawu YA, Kim HJ, Yun JH, et al. Enhancement of critical current density and strong vortex pinning in high entropy alloy superconductor Ta1/6Nb2/6Hf1/6Zr1/6Ti1/6 synthesized by spark plasma sintering. Acta Materialia. 2022;**232**: 117971

[22] Hong VTA, Jang H, Jung SG, Han Y, Kim JH, Hidayati R, et al. Probing superconducting gap of the high-entropy alloy Ta1/6Nb2/6Hf1/6Zr1/6Ti1/6 via Andreev reflection spectroscopy. Physical Review B. 2022;**106**(2):024504

[23] Jung SG, Han Y, Kim JH, Hidayati R, Rhyee JS, Lee JM, et al. High critical current density and high-tolerance superconductivity in high-entropy alloy thin films. Nature Communications. 2022;**13**(1):3373

[24] Zhang X, Winter N, Witteveen C, Moehl T, Xiao Y, Krogh F, et al. Preparation and characterization of high-entropy alloy (TaNb)1−x(ZrHfTi) x superconducting films. Physical Review Research. 2020;**2**(1):013375

[25] Pristáš G, Bačkai J, Orendáč M, Gabáni S, Košuth F, Kuzmiak M, et al. Superconductivity in medium-and high-entropy alloy thin films: Impact of thickness and external pressure. Physical Review B. 2023;**107**(2):024505

[26] Shu R, Zhang X, Rao SG, Le Febvrier A, Eklund P. Effects of alloying and deposition temperature on phase formation and superconducting properties of TiZrTaNb-based high entropy-alloy films. Applied Physics Letters. 2022;**120**(15):151901

[27] Marik S, Varghese M, Sajilesh KP, Singh D, Singh RP. Superconductivity in equimolar Nb-Re-Hf-Zr-Ti high entropy alloy. Journal of Alloys and Compounds. 2018;**769**:1059-1063

[28] Motla K, Meena PK, Singh D, Biswas PK, Hillier AD, Singh RP. Superconducting and normal-state properties of the high-entropy alloy Nb-Re-Hf-Zr-Ti investigated by muon spin relaxation and rotation. Physical Review B. 2022;**105**(14):144501

[29] Hattori T, Watanabe Y, Nishizaki T, Hiraoka K, Kakihara M, Hoshi K, et al. Metallurgy, superconductivity, and hardness of a new high-entropy alloy

superconductor Ti-Hf-Nb-Ta-Re. Journal of Alloys and Metallurgical Systems. 2023;**3**:100020

[30] Kitagawa J, Hoshi K, Kawasaki Y, Koga R, Mizuguchi Y, Nishizaki T. Superconductivity and hardness of the equiatomic high-entropy alloy HfMoNbTiZr. Journal of Alloys and Compounds. 2022;**924**:166473

[31] Zeng L, Zhan J, Boubeche M, Li K, Li L, Yu P, et al. Superconductivity in the bcc-type high-entropy alloy TiHfNbTaMo. Advanced Quantum Technologies. 2023;**6**:2300213

[32] Sobota P, Topolnicki R, Ossowski T, Pikula T, Pikul A, Idczak R. Superconductivity in the high-entropy alloy (NbTa)0.67(MoHfW)0.33. Physical Review B. 2022;**106**(18)

[33] Ishizu N, Kitagawa J. New high-entropy alloy superconductor Hf21Nb25Ti15V15Zr24. Results in Physics. 2019;**13**:102275

[34] Sarkar NK, Prajapat CL, Ghosh PS, Garg N, Babu PD, Wajhal S, et al. Investigations on superconductivity in an equi-atomic disordered Hf-Nb-Ta-Ti-V high entropy alloy. Intermetallics (Barking). 2022;**144**:107503

[35] Nelson WL, Chemey AT, Hertz M, Choi E, Graf DE, Latturner S, et al. Superconductivity in a uranium containing high entropy alloy. Scientific Reports. 2020;**10**(1):4717

[36] Von Rohr FO, Cava RJ. Isoelectronic substitutions and aluminium alloying in the Ta-Nb-Hf-Zr-Ti high-entropy alloy superconductor. Physical Review Materials. 2018;**2**(3):034801

[37] Krnel M, Jelen A, Vrtnik S, Luzar J, Gačnik D, Koželj P, et al. The effect of scandium on the structure, microstructure and superconductivity of equimolar Sc-Hf-Nb-Ta-Ti-Zr refractory high-entropy alloys. Materials. 2022;**15**(3):1122

[38] Harayama Y, Kitagawa J. Superconductivity in Al-Nb-Ti-V-Zr multicomponent alloy. Journal of Superconductivity and Novel Magnetism. 2021;**34**(11):2787-2794

[39] Zherebtsov S, Yurchenko N, Panina E, Tikhonovsky M, Stepanov N. Gum-like mechanical behavior of a partially ordered Al5Nb24Ti40V5Zr26 high entropy alloy. Intermetallics (Barking). 2020;**116**:106652

[40] Zhang X, Eklund P, Shu R. Superconductivity in (TaNb)1-x (ZrHfTi)xMoy high-entropy alloy films. Applied Physics Letters. 2023;**123**(5):051902

[41] Stolze K, Tao J, Von Rohr FO, Kong T, Cava RJ. Sc–Zr–Nb–Rh–Pd and Sc–Zr–Nb–Ta–Rh–Pd high-entropy alloy superconductors on a CsCl-type lattice. Chemistry of Materials. 2018;**30**(3):906-914

[42] Pan Y, He X, Zhou B, Strong D, Zhang J, Bin YH, et al. Elastic properties of a Sc–Zr–Nb–Ta–Rh–Pd high-entropy alloy superconductor. Materials Today Communications. 2022;**33**:104265

[43] Stolze K, Cevallos FA, Kong T, Cava RJ. High-entropy alloy superconductors on an α-Mn lattice. Journal of Materials Chemistry C. 2018;**6**(39):10441-10449

[44] Liu B, Wu J, Cui Y, Zhu Q, Xiao G, Wu S, et al. Structural evolution and superconductivity tuned by valence electron concentration in the Nb-Mo-Re-Ru-Rh high-entropy alloys. Journal of Materials Science and Technology. 2021;**85**:11-17

[45] Marik S, Motla K, Varghese M, Sajilesh KP, Singh D, Breard Y, et al. Superconductivity in a new hexagonal high-entropy alloy. Physical Review Materials. 2019;**3**(6):060602

[46] Lee YS, Cava RJ. Superconductivity in high and medium entropy alloys based on MoReRu. Physica C: Superconductivity and its Applications. 2019;**566**:1353520

[47] Liu B, Wu J, Cui Y, Zhu Q, Xiao G, Wu S, et al. Superconductivity in hexagonal Nb-Mo-Ru-Rh-Pd high-entropy alloys. Scripta Materialia. 2020; **182**:109-113

[48] Hirai D, Uematsu N, Saitoh K, Katayama N, Takenaka K. Superconductivity in high-entropy antimonide M1–x Pt x Sb (M = equimolar Ru, Rh, Pd, and Ir). Inorganic Chemistry. 2023;**62**(35):14207-14215

[49] Zhu Q, Xiao G, Cui Y, Yang W, Song S, Cao GH, et al. Structural transformation and superconductivity in carbon-added hexagonal high-entropy alloys. Journal of Alloys and Compounds. 2022;**909**:164700

[50] Liu B, Wu J, Cui Y, Zhu Q, Xiao G, Wu S, et al. Superconductivity and paramagnetism in Cr-containing tetragonal high-entropy alloys. Journal of Alloys and Compounds. 2021;**869**: 159293

[51] Xiao G, Zhu Q, Yang W, Cui Y, Song S, Cao GH, et al. Centrosymmetric to noncentrosymmetric structural transformation in a superconducting high-entropy alloy due to carbon addition. Science China Materials. 2023; **66**(1):257-263

[52] Xiao G, Yang W, Zhu Q, Song S, Cao GH, Ren Z. Superconductivity with large upper critical field in noncentrosymmetric Cr-bearing high-entropy alloys. Scripta Materialia. 2023; **223**:115099

[53] Motla K, Soni V, Meena PK, Singh RP. Boron based new high entropy alloy superconductor Mo0.11W0.11V0.11Re0.34B0.33. Superconductor Science and Technology. 2022;**35**(7):074002

[54] Kasem MR, Yamashita A, Goto Y, Matsuda TD, Mizuguchi Y. Synthesis of high-entropy-alloy-type superconductors (Fe,Co,Ni,Rh,Ir)Zr2 with tunable transition temperature. Journal of Materials Science. 2021;**56** (15):9499-9505

[55] Liu B, Wu J, Cui Y, Zhu Q, Xiao G, Wu S, et al. Superconductivity in cubic A15-type V–Nb–Mo–Ir–Pt high-entropy alloys. Frontiers of Physics. 2021;**9**

[56] Yamashita A, Matsuda TD, Mizuguchi Y. Synthesis of new high-entropy alloy-type Nb3 (Al, Sn, Ge, Ga, Si) superconductors. Journal of Alloys and Compounds. 2021;**868**:159233

[57] Nakahira Y, Kiyama R, Yamashita A, Itou H, Miura A, Moriyoshi C, et al. Tuning of upper critical field in a vanadium-based A15 superconductor by the compositionally-complex-alloy concept. Journal of Materials Science. 2022;**6**:15990-15998

[58] Mizuguchi Y, Kasem MR, Matsuda TD. Superconductivity in CuAl2-type Co0.2Ni0.1Cu0.1Rh0.3Ir0.3Zr2 with a high-entropy-alloy transition metal site. Materials Research Letters. 2021;**9**(3): 141-147

[59] Kasem MR, Yamashita A, Hatano T, Sakurai K, Oono-Hori N, Goto Y, et al. Anomalous broadening of specific heat jump at Tc in high-entropy-alloy-type

superconductor TrZr2. Superconductor Science and Technology. 2021;**34**:125001

[60] Pugliese GM, Tortora L, Tomassucci G, Kasem RM, Mizokawa T, Mizuguchi Y, et al. Possible local order in the high entropy TrZr2 superconductors. Journal of Physics and Chemistry of Solids. 2023; **174**:111154

[61] Fujita Y, Kinami K, Hanada Y, Nagao M, Miura A, Hirai S, et al. Growth and characterization of ROBiS2 high-entropy superconducting single crystals. ACS Omega. 2020;**5**(27):16819-16825

[62] Zeng L, Wang Z, Song J, Lin G, Guo R, Luo SC, et al. Discovery of the high-entropy carbide ceramic topological superconductor candidate (Ti0.2Zr0.2Nb0.2Hf0.2Ta0.2)C. Advanced Functional Materials. 2023;**33**: 2301929

[63] Shu H, Zhong W, Feng J, Zhao H, Yue B. Coexistence of superconductivity and ferromagnetism in high entropy carbide ceramics. arXiv:2307.16438

[64] Mizuguchi Y. Superconductivity in high-entropy-alloy telluride AgInSnPbBiTe5. Journal of the Physical Society of Japan. 2019;**88**:124708

[65] Kasem MR, Ishii R, Katase T, Miura O, Mizuguchi Y. Tuning of carrier concentration and superconductivity in high-entropy-alloy-type metal telluride (AgSnPbBi)(1-x)/4InxTe. Journal of Alloys and Compounds. 2022;**920**: 166013

[66] Ying T, Yu T, Shiah YS, Li C, Li J, Qi Y, et al. High-entropy van der Waals materials formed from mixed metal dichalcogenides, halides, and phosphorus trisulfides. Journal of the American Chemical Society. 2021;**143** (18):7042-7049

[67] Liu B, Yang W, Xiao G, Zhu Q, Song S, Cao GH, et al. High-entropy silicide superconductors with W5Si3-type structure. Physical Review Materials. 2023;7:014805

[68] Sogabe R, Goto Y, Mizuguchi Y. Superconductivity in REO0.5F0.5BiS2 with high-entropy-alloy-type blocking layers. Applied Physics Express. 2018;**11**: 053102

[69] Sogabe R, Goto Y, Abe T, Moriyoshi C, Kuroiwa Y, Miura A, et al. Improvement of superconducting properties by high mixing entropy at blocking layers in BiS2 -based superconductor REO0.5F0.5BiS2. Solid State Communications. 2019;**295**:43-49

[70] Fujita Y, Nagao M, Miura A, Urushihara D, Mizuguchi Y, Maruyama Y, et al. Effects of equivalent composition on superconducting properties of high-entropy REOBiS2 (RE = La, Ce, Pr, Nd, Sm, Gd) single crystals. Physica C: Superconductivity and its Applications. 2023;**608**:1354254

[71] Rost CM, Sachet E, Borman T, Moballegh A, Dickey EC, Hou D, et al. Entropy-stabilized oxides. Nature Communications. 2015;**6**(1):8485

[72] Oses C, Toher C, Curtarolo S. High-entropy ceramics. Nature Reviews Materials. Nature Research. 2020;**5**:295-309

[73] Akrami S, Edalati P, Fuji M, Edalati K. High-entropy ceramics: Review of principles, production and applications. Materials Science and Engineering R. 2021;**146**:100644

[74] Zhang RZ, Reece MJ. Review of high entropy ceramics: Design, synthesis, structure and properties. Journal of Materials Chemistry A. Royal Society of Chemistry. 2019;7:22148-22162

[75] Zhang J, Xu B, Xiong Y, Ma S, Wang Z, Wu Z, et al. Design high-entropy carbide ceramics from machine learning. npj Computational Materials. 2022;**8**(1):5

[76] Kasem MR, Nakahira Y, Yamaoka H, Matsumoto R, Yamashita A, Ishii H, et al. Robustness of superconductivity to external pressure in high-entropy-alloy-type metal telluride AgInSnPbBiTe5. Scientific Reports. 2022;**12**:7789

[77] Mizuguchi Y, Demura S, Deguchi K, Takano Y, Fujihisa H, Gotoh Y, et al. Superconductivity in novel BiS 2-based layered superconductor LaO1-xFxBiS2. Journal of the Physical Society of Japan. 2012;**81**:114725

[78] Bérardan D, Franger S, Dragoe D, Meena AK, Dragoe N. Colossal dielectric constant in high entropy oxides. Physica Status Solidi - Rapid Research Letters. 2016;**10**(4):328-333

[79] Sarkar A, Wang Q, Schiele A, Chellali MR, Bhattacharya SS, Wang D, et al. High-entropy oxides: Fundamental aspects and electrochemical properties. Advanced Materials. 2019;**31**:1806236

[80] Li H, Zhou Y, Liang Z, Ning H, Fu X, Xu Z, et al. High-entropy oxides: Advanced research on electrical properties. Coatings. 2021;**11**:628

[81] Brahlek M, Gazda M, Keppens V, Mazza AR, McCormack SJ, Mielewczyk-Gryń A, et al. What is in a name: Defining 'high entropy' oxides. APL Materials. 2022;**10**:110902

[82] Aamlid SS, Oudah M, Rottler J, Hallas AM. Understanding the role of entropy in high entropy oxides. Journal of the American Chemical Society. American Chemical Society. 2023;**145**: 5991-6006

[83] Mazza AR, Gao X, Rossi DJ, Musico BL, Valentine TW, Kennedy Z, et al. Searching for superconductivity in high entropy oxide Ruddlesden–Popper cuprate films. Journal of Vacuum Science & Technology A. 2022;**40**: 013404

[84] Wang K, Hou Q, Pal A, Wu H, Si J, Chen J, et al. Structural and physical properties of high-entropy REBa2Cu3O7-δ oxide superconductors. Journal of Superconductivity and Novel Magnetism. 2021;**34**(5):1379-1385

[85] Yamashita A, Shukunami Y, Mizuguchi Y. Improvement of critical current density of RE Ba2Cu3O7-δ by increase in configurational entropy of mixing. Royal Society Open Science. 2022;**9**(3):211874

[86] Suzuki Y, Nagao M, Maruyama Y, Watauchi S, Tanaka I. Growth of REBa2Cu3Ox single-crystal whiskers utilizing the concept of high-entropy alloys. Japanese Journal of Applied Physics. 2023;**62**:033001

[87] Yamashita A, Hashimoto K, Suzuki S, Nakanishi Y, Miyata Y, Maeda T, et al. Fabrication of high-entropy REBa2Cu3O7– δ thin films by pulsed laser deposition. Japanese Journal of Applied Physics. 2022;**61**(5):050905

[88] Chen J, Huang R, Zhou X, Zhou D, Li M, Bai C, et al. Nucleation and epitaxy growth of high-entropy REBa2Cu3O7–δ (RE= Y, Dy, Gd, Sm, Eu) thin films by metal organic deposition. Journal of Rare Earths. 2023;**41**(7):1091-1098

[89] Masuda H, Ishii R, Kita R, Miura O. Superconducting properties of high-entropy type RE123 thin films by fluorine-free MOD method. IEEE Transactions on Applied Superconductivity. 2023;**33**(5):7200203

[90] Cayado P, Grünewald L, Erbe M, Hänisch J, Gerthsen D, Holzapfel B.

Critical current density improvement in CSD-grown high-entropy REBa2Cu3O7−δ films. RSC Advances. 2022;**12**(44):28831-28842

[91] Pryanichnikov S, Vidmid' L, Titova S. High-entropy superconducting oxides (Y,Nd,Eu,Sm,Ho)Ba2Cu3Oy with different oxygen contents. Journal of Superconductivity and Novel Magnetism. 2023;**36**(3):871-875

[92] Grünewald L, Cayado P, Erbe M, Hänisch J, Holzapfel B, Gerthsen D. Analytical electron microscopy study of the composition of $BaHfO_3$ nanoparticles in REBCO films: The influence of rare-earth ionic radii and REBCO composition. Materials Advances. 2023; **4**:6507-6521